By William R. Gray   Photographed by Gordon W. Gahan
Prepared by the Special Publications Division
National Geographic Society, Washington, D.C.

Voyages to Paradise:

Exploring in the Wake of Captain Cook

VOYAGES TO PARADISE:
EXPLORING IN THE WAKE OF CAPTAIN COOK
BY WILLIAM R. GRAY

Photographed by GORDON W. GAHAN

Published by
  The National Geographic Society
  GILBERT M. GROSVENOR, *President*
  MELVIN M. PAYNE, *Chairman of the Board*
  OWEN R. ANDERSON, *Executive Vice President*
  ROBERT L. BREEDEN, *Vice President, Publications and Educational Media*

Prepared by
  The Special Publications Division
  DONALD J. CRUMP, *Editor*
  Philip B. SILCOTT, *Associate Editor*
  WILLIAM L. ALLEN, WILLIAM R. GRAY, *Senior Editors*

  MARY ANN HARRELL, *Managing Editor*
  THOMAS B. POWELL III, *Picture Editor*
  JODY BOLT, *Art Director*
  BONNIE S. LAWRENCE, *Assistant to the Picture Editor*
  TONI EUGENE, BONNIE S. LAWRENCE, *Researchers*
  DONNA B. KERFOOT, *Assistant Researcher*

Illustrations and Design
  SUEZ B. KEHL, *Assistant Art Director*
  CYNTHIA B. SCUDDER, *Assistant Designer*
  CONNIE B. BOLTZ, *Assistant for Calligraphy*
  RICHARD FLETCHER, *Design Assistant*
  JOHN D. GARST, JR., LISA BIGANZOLI, PATRICIA K. CANTLAY, MARGARET
    DEANE GRAY, MARK SEIDLER, *Map Research, Design, and Production*
  LESLIE B. ALLEN, MICHAEL C. BLUMENTHAL, TONI EUGENE, JUDITH A.
    FOLKENBERG, WILLIAM R. GRAY, KATHLEEN F. TETER, *Picture Legend Writers*

Engraving, Printing, and Product Manufacture
  ROBERT W. MESSER, *Manager*
  GEORGE V. WHITE, *Production Manager*
  RICHARD A. MCCLURE, *Production Project Manager*
  MARK R. DUNLEVY, RAJA D. MURSHED, CHRISTINE A. ROBERTS, DAVID V.
    SHOWERS, GREGORY STORER, *Assistant Production Managers*
  SUSAN M. OEHLER, *Production Staff Assistant*
  DEBRA A. ANTONINI, NANCY F. BERRY, PAMELA A. BLACK, BARBARA BRICKS,
    JANE H. BUXTON, MARY ELIZABETH DAVIS, ROSAMUND GARNER, VICTORIA
    D. GARRETT, NANCY J. HARVEY, SUZANNE J. JACOBSON, ARTEMIS S.
    LAMPATHAKIS, VIRGINIA A. MCCOY, MERRICK P. MURDOCK, CLEO PETROFF,
    MARCIA ROBINSON, CAROL A. ROCHELEAU, KATHERYN M. SLOCUM, JENNY
    TAKACS, PHYLLIS C. WATT, *Staff Assistants*
JOLENE M. BLOZIS, *Index*

Unspoiled vistas still mark James Cook's track through the Pacific Ocean. At right, a rainbow shimmers in the valley of the Vaitoto River, on Tahiti. Pages 2-3: Below cloud-dimmed Mont Panié, net fishermen work at sunrise in a cove on the eastern coast of New Caledonia, discovered and named by Cook on his second voyage, in 1774. Page 1: Girls of Aitutaki, in the southern Cook Islands, wade on a breeze-swept beach. Hardcover: Cook's coat of arms, awarded after his death by George III, places the sphere of the Pacific between two polar stars; the wreath on the crest combines the classic laurel of victory with a frond of tropical palm.

Contents

Prologue: "Farther Than
Any Other Man Has Been"   6

1   The Early Years:
"A Bloody Good Seaman"   10

2   The First Voyage:
"The Entrancing
Cloud-hung Heights
of Tahiti"   32

3   The First Voyage:
"Into Uncharted Seas"   66

4   The Second Voyage:
"From Icebound Seas to
Sun-blessed Islands"   102

5   The Third Voyage:
"The Call of My Country for
More Active Service"   140

6   The Third Voyage:
"The Frozen Secrets of the
Artic"   166

7   The Third Voyage:
"The Man They
Worshipped as a God"   190

Epilogue: "A Man of
Unparalleled Success"   210

Acknowledgments
and Illustrations Credits   212

Additional Reading   212

A Note on Quotations   212

Index   212

Prologue:
"Farther Than Any Other Man Has Been"

Rampaging gales bullied the blue Pacific, pushing violent breakers onto jagged coral heads. Trapped by wind and current in a maze of reefs off New Caledonia, the crew of the little sailing ship *Resolution* faced a perilous night. At any moment she could be driven onto the rocks and shattered. "...I realy think our situation was to be envyed by very few except the Thief who has got the Halter about his Neck," one man recalled.

Throughout that dark night, the heart-stopping cry of "Breakers ahead!" rent the tropic air. The captain, a veteran of exploring the Pacific Ocean, snapped orders that kept his ship tacking away from danger. In the morning he wrote in his journal: "Day-light shewed ... that we had spent the night in the most eminent danger havᵍ had shoals and breakers continually under our lee at a very little distance from us."

Brilliant seamanship and a brilliantly disciplined crew had rescued the ship from disaster. Any ordinary commander would have set sail immediately for the open sea, thankful to escape that tangle of reefs.

But this was no ordinary commander. James Cook, a superb sailor, also possessed a wide-ranging scientific curiosity. He had approached this dangerous shoreline because he had noted unusual objects spiking the land—columns of rock, or bizarre trees. "I was now almost tired of a Coast I could no longer explore but at the risk of loosing the ship and ruining the whole Voyage, but I was determined not to leave it till I was satisfied what sort of trees those were...."

So he edged along through treacherous shallows, to anchor near an islet with a cluster of the trees. With his expedition's botanists, Cook went ashore and marveled at one of the strangest plants in the South Pacific—*Araucaria columnaris,* once *cookii*. It soars as tall as 200 feet, but its branches rarely exceed 6.

For the moment Cook was satisfied—he

CAPTAIN JAMES COOK. PORTRAIT FROM LIFE BY NATHANIEL DANCE, 1776; NATIONAL MARITIME MUSEUM, GREENWICH, ENGLAND

had investigated those puzzling trees, he and his companions had sampled the island's plant and animal life, and he had made his usual accurate observations of latitude and longitude. After gathering a few last botanical specimens, Cook turned the *Resolution* toward the south.

A soft-spoken man of science, a strong-willed man of the sea, James Cook, in three globe-girdling voyages from 1768 to 1779, explored more of the Pacific Ocean than any man before him. In so doing, he discovered a myriad of untracked islands, he immeasurably expanded many fields of knowledge, and he stirred the imaginations of people around the world.

Born in humble circumstances in 1728, Cook ascended to the forefront of public and intellectual life in England. Early in his career, he evinced characteristics that aided his rise: dogged perseverance, navigational expertise, uncanny leadership skills, marked resourcefulness and ability to make decisions, and a general competence in many areas of learning.

"James Cook was one of the great men of the 18th century," historian Michael E. Hoare told me in Wellington, New Zealand. Dr. Hoare's academic honors include a term of study as a "James Cook Fellow."

"Cook came to the fore at an extremely important moment in history," he continued. "Technological, scientific, and medical discoveries were burgeoning — many beneficial to long voyages of exploration. The 18th century was the Age of Reason, in which a desire for knowledge of the natural world was rekindled. It was also a time of romantic vision, when the thought of discovering new lands and peoples was appealing. In addition, the British Empire was poised for great expansion. The Seven Years War between England and France had ended in 1763. The victory propelled Britain into a period of extraordinary activity. Captain Cook's successes, I think, were caused by the conjunction of an intelligent, practical, and capable man with the situation and the times that called for one.

"We cannot, of course, overlook the factor of luck. And Cook was fortunate to have excellent crews and some of the world's most respected scientists on his expeditions. But he was truly a remarkable man and one whose significance will never fade in the study of history."

Becalmed somewhere at sea, H.M.S. Resolution *—Cook's flagship on his second and third great voyages —awaits a favoring wind. Midshipman Henry Roberts, R.N., did this watercolor study some time on the second. Experts consider it the best portrayal of the type of vessel Cook chose for exploration: the cat-built bark of the North Sea coal trade. A "cat" rolled heavily in a seaway, and made about eight knots at best, but her sturdiness made her a supremely reliable ship.*

With veteran National Geographic photographer Gordon Gahan, I pursued the wake of this remarkable man for nearly a year. Our travels took us about the vast Pacific Ocean — from the northern tip of Alaska to the southern tip of Tasmania, and to most of the island groups in between. We traveled to England, to the moors of North Yorkshire where Cook spent his childhood, to the North Sea ports where he learned the crafts of sail; to the Maritime Provinces of Canada where he matured as a commander and as a scientific observer. We journeyed to Tahiti and New Zealand — two of Cook's favorite refuges; to Australia; to the many South Pacific paradises that he discovered; to the ice in Bering Strait; and finally to Hawaii, where he met his death.

With every mile, with every landfall, I marveled anew at the accomplishments of James Cook. Although Ferdinand Magellan had crossed the Pacific nearly 250 years before Cook, and the Portuguese, Spanish, Dutch, French, and English had explored parts of it, the Pacific Ocean in 1768 remained an enormous mystery. Few reliable charts existed.

"To me, the thought of Cook and his men setting off into the total unknown is quite terrifying," Rear Admiral D. W. Haslam, Hydrographer of the Royal Navy, told me in his office in Taunton, England. "Month after month — for years at a time — Cook navigated uncharted waters, often in abhorrent conditions of wind and weather. It's incredible! He surely had a seventh sense for the sea. He safely guided those small wooden ships — not much bigger than Thames River barges — through frozen seas packed with icebergs and through tropic seas filled with coral. He faced all the conditions dauntlessly."

Even today he is a legend throughout the Pacific. One venerable islander from Tonga, whose Polynesian forebears were among the best navigators in the world, said flatly, "Besides Jesus Christ, Captain Cook was the greatest man who ever lived."

Cook's name graces a picturesque town in Australia, the highest mountain in New Zealand, a jewel-like strand of islands in the central Pacific, a mountain-rimmed inlet in Alaska, scores of other geographic features. But that came as a tribute from those who followed him. He named his discoveries after his patrons in England, his crewmen, even his ships. Cook was modest, hardworking, taciturn.

A large, rawboned man, he had a certain stolidity of character. He strove to be just with his men — although at times he revealed a violent temper. In an era when impressment was the surest way to form a ship's company, men willingly sailed with Cook. Although some called him "the despot," most respected and even loved him.

The same held true for the islanders he met. Captain Cook treated them as fairly as he did his men, and generally they returned his kindness with veneration. Problems, of course, arose, but usually from misunderstandings in the collision of two disparate cultures.

Cook kept detailed journals of his voyages, with observations on places and peoples and interpretations of various events. Only by implication, as a rule, did he disclose his personality. Occasionally, however, the sheer excitement of what he had accomplished inspired him, and his pride became evident. On his second voyage, after sailing far south of the Antarctic Circle, Cook revealed this of himself: "ambition leads me not only farther than any other man has been before me, but as far as I think it possible for man to go. . . ."

That indeed illuminates his character. To survive the perils of the Pacific Ocean — storm waves that dwarf a ship, coral reefs that crush a hull, sun that blisters the skin, cold that transforms rigging into flesh-shredding cables of ice — James Cook had to be ambitious, and to know himself well.

Most of our intimate knowledge of Cook comes from those who wrote about him: members of his expeditions, friends and associates in England, and biographers, especially John Cawte Beaglehole of New Zealand. Professor Beaglehole, who died in 1971, devoted his long career to the study of James Cook. He edited and published the journals, and he wrote the definitive study of the life. In "those few elected spirits, such as Cook," mused Dr. Beaglehole, "is the complete equipment of genius, and fortune coincides with their appearance, and the face of the world is changed."

As we traced him across the Pacific, Gordon and I quickly discovered the awesome degree to which Captain James Cook had changed the face of the world.

North Atlantic
Ocean

NORTH
AMERICA

St. Lawrence

Newfoundland

• Quebec

• Louisbourg
— *Cape Breton Island*
Halifax
— *Nova Scotia*

0   500
STATUTE MILES

*North
Sea*

— Staithes
• Whitby
England

London —
English Channel

France

EUROPE

The Early Years:
"A Bloody Good Seaman"

On a chill October day in 1728, a son was born to a simple farm couple living in a two-room thatched cottage in the north of England—an infant named James after his father, an industrious and honest farm worker from Scotland. On November 3, James and Grace Cook took their week-old son to the parish church of St. Cuthbert in the hamlet of Marton-in-Cleveland to be baptized. Thus was James Cook — explorer, circumnavigator, Fellow of the Royal Society, honored visitor at the court of King George III, and post captain in the Royal Navy—presented to the world.

His boyhood in Marton, then a pastoral village and now a busy suburb of Middlesbrough, was generally uneventful. He was a robust and healthy lad, although four of his seven brothers and sisters died in early childhood. Dame Walker, wife of a local yeoman farmer, apparently took an interest in him; she taught him the alphabet and a bit of reading in exchange for chores around her farm. The elder James Cook instilled such north-country virtues as sobriety, diligence, and hard work.

"The Cooks also regularly attended church," the Reverend Ronald M. Firth, the current vicar at St. Cuthbert's, told me. A brisk autumn wind chased fallen leaves around moldering gravestones as we strolled toward the church. Built solidly of local brownstone, it stands under a canopy of stately oaks and elms. "It dates back to the 1150s," said the vicar as he guided me through a nave filled with monuments commemorating local worthies. "Over there, in the north transept, is where the Cook family always sat. They kept that pew the entire time they were in Marton."

That time ended in 1736, however, when James was eight years old. His father accepted the respectable position of hind, or foreman, at Aireyholme Farm near Great Ayton, a pleasant community nestled by the Cleveland Hills of North Yorkshire. Thomas Skottowe, Lord of the Manor of Ayton, apparently saw promise in young James; in fact, he became the first of the significant patrons who, at pivotal moments, favored the career of James Cook. Mr. Skottowe paid his fees at the little Postgate School in Ayton. Although no shining scholar, James reportedly excelled in arithmetic.

Early on, he showed traits of leadership—and stubbornness. According to a local tradition recorded in 1808, he would never give up a plan of his own to please his playmates. Often he would go out in the evening with other village boys to look for birds' nests. He would argue, "strenuously," for a particular location: "sometimes . . . with such inflexible earnestness, as to be deserted by the greater part of his companions."

During his years at Ayton, young James surely developed his fascination with the sea — and it must have drawn him. Rising directly behind Aireyholme Farm is a hill called Roseberry Topping, its once-rounded summit now canted by mining to resemble a gigantic wave frozen at the point of breaking. I can imagine the farm lad, his chores complete on a lazy summer evening, scampering up the hill, stretching out on its rugged top, and gazing out across the miles to the North Sea. Large ships might have been beating against the wind, moving in the slow graceful fashion that only sail can give. Perhaps a roak — a dense sea mist — might have crawled inland at dusk, bringing the thick, musty, irresistible smells of the sea.

At 16 or 17, James left his parents' home and moved to the tiny seaport of Staithes. It was Thomas Skottowe, no doubt, who obtained him an informal trial apprenticeship in the grocery shop and haberdashery of his friend William Sanderson. A fishing port, Staithes crouches at the base of a crumbling cliff of clay and sand and gravel, and the gray tides of the North Sea

PRECEDING PAGES: *Shrieking gulls pierce the mist above Staithes on England's blustery Yorkshire coast. Here, where men brave the icy waves of the North Sea in open boats, James Cook first felt salt spray and the lure of life at sea.*

slowly erode it away. Indeed, the site of Sanderson's shop now lies some 20 yards off the shoreline. The life of a grocer's boy in this quiet setting clearly did not suit James, for within 18 months he had moved on. But in Staithes, he truly became accustomed to the sea.

It would be hard not to: The sea dominates everything. On a drizzly autumn evening, Gordon and I walked the narrow, twisting streets of Staithes. Fog cloaked the little place, which still retains much of its 18th-century flavor. For several blocks we passed only shuttered storefronts—and saw no one. Finally we came to the harbor. The smells of fish, rotting seaweed, and the sea itself pervaded all. Three boys silently and monotonously tossed rocks into the surf. "Living in a place like this," Gordon said, "I can see how one could become entranced by the sea."

During his year and a half behind Sanderson's counter, Cook constantly saw ships coursing offshore and heard the tales of fishermen. His yearning for the sea must have grown, and Sanderson must have agreed to let him move to nearby Whitby, a teeming North Sea port.

Among its most respected merchants and shipowners were the brothers Walker, John and Henry, and they needed hardworking apprentices. James moved into the home of John Walker—another of those men who had major significance in Cook's life. He was a Quaker, a fair and honest man, who encouraged his young apprentice to learn the rudiments of navigation and the use of shipboard equipment. Diligently, over the next few winters, James studied late at night, with a candle and a small table provided by the housekeeper, Mary Prowd, who seemed to dote on Cook.

In the summer of 1746, Cook signed aboard one of the Walkers' ships as "servant" and sailed from the harbor of Whitby. Like most shipowners in this region, the Walkers were engaged in the coal trade. Their colliers would sail north, take on coal from Newcastle, then go to London and sell it. During a good season — from the end of February to just before Christmas — each vessel might make ten round trips. It was hard work, especially for the apprentices. They were expected to be the first on deck, the first up the rigging, the first to man the capstan and windlass. Those who took well to the work might hope for advancement; those who didn't earned the scorn of their comrades, and the grubbiest tasks. Cook advanced quickly.

A bright and earnest young man, Cook watched the master handle the ship, learned the proper commands and their results, asked questions. He also experienced some of the most challenging waters in the world. It has been said for centuries that the North Sea is the nursery of seamen, and it is true. Captain Alan Villiers — a veteran sailor and author, and probably the last man to command a full-rigged ship on a voyage around the world—has commented: "the North Sea, with its hazards of sandbanks and racing tides, lee shores, poorly marked roadsteads and indifferent harbors difficult to approach, frequent gales and sudden shifts of wind, provided an excellent and most testing sailing ground."

After his earliest voyages, Cook sailed on the Walkers' *Freelove* on September 29, 1747 — and probably got his first sight of London. He was certainly learning the ways of his ship. She was a three-master, square-rigged; 106 feet long and 27 feet in the beam, broad of hull to carry several hundred tons of cargo. She was somewhat squat, and had a flat bottom that let her rest on a mudflat at low tide — she was perfect for sailing in bad weather and in shallow seas. Cook came to trust such "cat-built barks," the colliers of Whitby, famed in their day as "built remarkably strong."

Between trips, Cook stayed with John Walker in Whitby. In his spare time he probably wandered about to visit other ships and talk with sailors.

Gordon and I arrived in Whitby on a Saturday afternoon in the midst of an October rainstorm driven by a blustery wind. Despite the weather, the

harbor was bustling with activity — a clear contrast to the solemnity of Staithes. Cook would have found it so, for in the 18th century Whitby was at the pinnacle of its shipbuilding life. To escape the rain, Gordon and I dodged sprawling puddles and tumbled into the Plough, a pleasantly traditional pub in the center of town. Over platters of steaming shepherd's pie and pints of hearty English ale, hand-pumped from wooden casks, we dried out in the cozy familiarity of the pub. Fishermen, shopkeepers, sailors, families, and tourists ebbed into the Plough, but the conversations inevitably revolved around the topics of the weather, the sea, and the financial timber of the times — subjects discussed over the ale for centuries. I could picture young James Cook sitting in a corner, sipping ale or tangy cider and listening to boisterous and boastful stories.

History repeats itself on Aireyholme Farm: Where James Cook once tended livestock, Mark Phalp, 15, and his dog Pod stride between bleating lambs and centuries-old cowsheds. Mark shares his famous predecessor's desire: to see the world. Cook's baptismal record (above) survives in the parish church at Marton, but 8-year-old James moved with his family to Aireyholme when his father became farm foreman. Chores and several years of formal education at the village school occupied the lad until he left home in his mid-teens. Transported from Yorkshire to Australia as a memorial, the home of Cook's parents now stands in Melbourne's Fitzroy Gardens.

PRECEDING PAGES: *Weathered sandstone walls stitch patchwork fields in a Yorkshire dale. The son of a laborer, Cook spent his boyhood on such a farm, and in that peaceful setting learned the necessity of hard work and the virtue of diligence.*

Quaint and crowded,
Staithes defies the slate-
gray waters of the North Sea.
Hired at a small grocery-and-
notions shop on the seafront,
Cook glimpsed fishing boats
challenging the waves.
Grounded at low tide,
traditional craft, called cobles,
continue to serve the
doughty men of Staithes.

Cook sailed repeatedly as servant. Then in April of 1749, his apprentice-ship over, his name first appears on the muster rolls as "seaman." His progress in navigation and leadership must have been extraordinary, for John Walker soon promoted him to mate and then, early in 1755, offered him the honorable position of master. For a Yorkshire farm boy of humble birth, this was indeed a distinction. Cook surely considered the proposal seriously. Despite the many dangers of the trade, it offered him a secure life with a stable and prosperous firm — and the prospect of one day becoming a shipowner himself. But for James Cook, age 27, this was not enough.

On June 17, 1755, he made a momentous decision. In Wapping, a dockside district of London, he volunteered for the Royal Navy as an able seaman. To friends and shipmates it must have seemed a strange choice. Conditions and pay in the navy were inferior to those in the merchant service: the food terrible, health conditions abhorrent, discipline enforced by the whip. Most of the officers gained their rank by virtue of their social status, while the lower decks were often full of hooligans and drunkards forcibly impressed into service. "Manned by violence and maintained by cruelty" — so a respected veteran admiral described the fleet.

Into this unsavory atmosphere came James Cook, impelled perhaps by patriotic fervor. Britain and France were maneuvering toward war, a con-flict that would be called the Seven Years War in Europe, the French and Indian War in America. In it, the British navy would play a decisive role and improve its seamy reputation. Some capable men would rise quickly and gain the notice of the powerful. James Cook would be one of these men.

He was assigned immediately to H.M.S. *Eagle,* a 60-gun warship. Within weeks her captain promoted him to master's mate. In August the *Eagle* sailed down the Thames to take up station off the coast of France. Hostilities had already begun, though war was not declared until May 1756.

Before the *Eagle* engaged the enemy, however, she acquired a new captain, a young gentleman of old Yorkshire stock. At 33, Hugh Palliser already had served for 20 years in the navy, and his future was bright. Cook and Palliser, two Yorkshiremen, gained each other's lasting respect.

For nearly two years the *Eagle* cruised the English Channel and its western approaches, occasionally capturing French ships of commerce and war. Her biggest prize, and one that brought praise from the Admiralty, was the East Indies merchantman *Duc d'Aquitaine,* armed with 50 heavy guns and won only after a fierce hour-long battle.

At about this time, Captain Palliser, urged on by Thomas Skottowe and John Walker, recommended his mate for the position of master, "a station that he was well qualified to discharge with ability and credit." A warship's master had many essential duties, with special responsibility for navigation; in that age a man of modest social position and great professional skill could expect no higher naval rank. Cook passed the written and oral examination, a stiff one, and received his master's warrant on June 29, 1757.

After a short stint on patrol off Scotland, Cook was appointed master of the new warship *Pembroke,* and sailed with her to North America under her benevolent captain, John Simcoe. Canada would be a major theater of war, and a rich prize of battle. Two fortified cities — Louisbourg on the Atlantic coast of Cape Breton Island, and Quebec far inland on the St. Lawrence River—formed the bulwarks of New France, and blocked British expansion.

"You must understand that the Seven Years War was crucial, a great

Golden lamplight still welcomes sailors to the port of Whitby, Cook's home for nine years. Apprenticed to a shipowner, Cook mastered inshore sailing by signing on the sturdy coal-trade vessels that plied a treacherous coastal route to London.

turning point in history." Ken Donovan, a staff historian at the Fortress of Louisbourg National Historic Park, was steering me through time. "It was the culmination of a long economic and military rivalry between England and France. The two powers had this new and vast tract of land — North America — on which to compete. Its resources seemed limitless, and whoever controlled it, seemingly, would control the future."

"The fisheries were highly important," added Sandy Balcom, another staff historian. "Before the war the yearly cod catch from Louisbourg was three times as valuable as the fur trade from Quebec. The French built the fortress here to protect this rich resource, important even today."

The British had to take Louisbourg, key defense of the St. Lawrence, before moving against Quebec. The *Pembroke* joined a fleet of 157 vessels for

*O*nly *wild flowers and high grass invade the courtyard of the King's Bastion Barracks as marines drill in the reconstructed fortress of Louisbourg, once the stronghold guarding France's North American colonies. When Cook joined the Royal Navy in 1755, war threatened between France and England; after a few years' duty in home waters, he participated in the siege of Louisbourg that opened Canada to British conquest. Jutting from a rocky peninsula of Cape Breton Island, the fortress commanded the Gulf of St. Lawrence, sea route to Quebec, capital of New France. Britain, resentful of harassment of its ships by privateers from Louisbourg, and long envious of French wealth in North America, attacked the stronghold in 1758. After it fell, Cook helped chart the tricky channel of the St. Lawrence River so the English fleet could safely approach, then subdue Quebec.*

the siege that began June 8, 1758, under Gen. James Wolfe. His guns battered down the defenses, and on July 26 the governor surrendered.

The next day a chance encounter proved a turning point in Cook's career. Walking on a beach near Louisbourg, he spotted a man stooping low to squint along a square table perched on a tripod. The man would record notes and then move his apparatus to another spot. Curious, Cook struck up a conversation. Surveyor Samuel Holland was making a detailed plan of the harbor, using a portable plane table. Cook expressed an interest in learning to use it, Captain Simcoe approved, and Holland "appointed the next day in order to make him acquainted with the whole process. . . ."

Cook and Holland were inseparable for the next few weeks. By November Cook had completed a chart of Gaspé Bay—the first of hundreds

he would make—that was published the next year in London. In this quiet way, James Cook took up an occupation that would absorb him for a lifetime.

Britain's admirals and generals in Canada agreed that the season was too late to risk an attack on Quebec, so the fleet anchored for the winter in the commodious harbor of Halifax, Nova Scotia. Almost daily, Cook, Holland, and Simcoe gathered to work on charts and to discuss techniques of mapping. The captain encouraged Cook to learn "Spherical Trigonometry, with the practical part of Astronomy." In a practical way, the three compiled a provisional chart of the St. Lawrence—unless the British ships could ascend it in good time and good order, their whole New World campaign might fail.

Finally, next May, the ice broke up and the ships moved tentatively upstream. Instead of the season's normal westerly, they had a favoring wind from the northeast. Much of this great river was easily navigable—given a fair wind—but in the 120 miles below Quebec City it became perverse. In those reaches, shifting shoals and banks altered the channel, particularly in La Traverse, a treacherous region off the Ile d'Orléans—almost within sight of the city. From a hilltop on the island, I gazed out at this notorious passage, where the old channel veered from the north shore to the south. Under the surface, which seemed placid enough, a maze of mudbanks still requires frequent dredging.

"D__ me, if there are not a thousand places in the Thames fifty times more hazardous than this. . . ." So one salty merchant master commented after Cook and other British masters had sounded, marked, and charted the channel: Not one ship of 168 was lost. On June 27 the fleet lay anchored beyond range of the clifftop guns of Quebec.

Skirmishes, feints, and raids occupied the opposing forces for nearly three months, until General Wolfe perfected a plan for capturing the city. In matter-of-fact style, Cook recorded this famous exploit in his ship's log: ". . . at midnight all the Row Boats in the fleet made a faint to Land at Beauport . . . to favor the Landing of the Troops above the Town on the north Shoar . . . the English Army Command^d by Genl Wolf, attacked the french under the Com^d of Genl Montcalm in the field of Aberham behind Quebec, and Tottally Defeated them. . . ."

Entering triumphantly on September 18, the redcoats settled down in Quebec for the winter. Cook was transferred to H.M.S. *Northumberland*, captained by Alexander, Lord Colville, who assumed command of the fleet. Before the St. Lawrence iced over, the ships sailed for Halifax. Cook would

spend two winters there, with fair-weather duty on the coast of Nova Scotia. As usual, he busied himself with charts and soundings, and compiled sailing directions for areas as remote as the West and the East Indies. He obviously impressed his commander, for on January 19, 1761, Colville did something most unusual—he ordered a bonus to Cook of £50, almost eight months' regular pay, "in consideration of his indefatigable Industry in making himself Master of the Pilotage of the River Saint Lawrence...."

After more brief service off Newfoundland, the *Northumberland* returned to London in October 1762. Cook submitted his carefully prepared papers to the Admiralty and left the ship he had served on for three years, undoubtedly wondering what orders the navy would have for him next. Before the war officially ended, Rear Admiral Lord Colville posted a letter to

the Admiralty: "... I beg leave to inform their Lordships, that from my Experience of Mr Cook's Genius and Capacity, I think him well qualified for the Work he has performed, and for greater Undertakings of the same kind."

Such "greater Undertakings" soon materialized. Their Lordships duly chose Cook to survey the rocky, storm-lashed coasts and straits and isles of eastern Canada.

Characteristically, he went about organizing the equipment he would need. Uncharacteristically, he was more than two weeks late joining his ship. James Cook, the sober, orderly, and hardworking master in the Royal Navy, had fallen in love and been married.

Louisbourg lives again as a Canadian National Historic Park. From an inn doorway, Denise LeBlanc invites guests to the 18th century; costumed children enhance the mood. Louisbourg thrived as harbor, fishing port, and home to 5,000. The British leveled the fortifications; Canada began rebuilding in 1961.

Of his courtship we know nothing. But on December 21, 1762, Cook, age 34, escorted Elizabeth Batts, age 21, to the parish church of St. Margaret's in the village of Barking, and the two were joined in matrimony. They took lodgings in Shadwell, near the Thames; and in March James began preparing to leave for the New World. Because Mrs. Cook later destroyed all personal papers, we know little of their married life—except that it was a loyal and happy one in spite of many long separations.

In Canadian waters, Cook's first chore was to map the isles of St. Pierre and Miquelon, off southern Newfoundland. Losing her immense Canadian domain under the peace treaty, France was to keep these bits of land as a base for her North Atlantic fishery. Before handing them over, Britain wanted them charted, accurately. Cook worked with his usual diligence; by August the islands were returned to France. And to France they belong to this day. St. Pierre has the essence of any French coastal village — a moody, earthy charm. I bought French croissants with French francs, enjoyed excellent French wines with my meals, and even had the same difficulties with aloof taxi drivers that I've had in Paris.

Back in Newfoundland, Cook enjoyed a surprise: the command of his own ship, a 68-ton schooner called the *Grenville.* He immediately sailed north, to work the coast of Labrador until the end of the season.

He reached home in late November and discovered that his young wife had given him a son, named James. He bought a small house for his family in the village of Mile End Old Town on the outskirts of London, and settled happily to work — preparing final copies of his new charts, soundings, and sailing directions. He had submitted his preliminary work direct to the Lords of the Admiralty and was "convinced it was well received." Personal acquaintance with their Lordships would serve him well in later years.

Now Hugh Palliser was appointed governor of Newfoundland; and over the next four summers, as Cook continued his surveys, their friendship deepened. Before the winter of 1767 Cook covered a third of the island's 6,000-mile coastline. It was tedious work on a coast infamous for its rocks and ice and shoals, its haze and fog, its unpredictable gales.

I learned something of its nature when I sailed with a group of cod fishermen from the enchanting coastal village of Brigus South.

"Aye, the sea is still the way to earn a living here," said Greg Doyle in the rolling brogue of a "Newfie." "Has been for many years and will be for many more." A ruddy man of 42, Greg steered his wooden skiff along the austere coast, pounded by a heavy surge. A thick gray mist deepened the gray of the rocks and the sea, but seemed to brighten the scattered patches of green grass. "We build these boats the way they were built in Captain Cook's time," Greg continued, "and we build them to last." On the gunwales were grooves where ropes had rubbed the wood for years, and a small, jury-rigged engine was the only concession to modernization.

Bracing myself against the ceaseless rolling, I stood in the bow and peered ahead into the dank mist. I sensed something of the tension Cook must have felt along this shore. Craggy headlands and barely submerged rocks slipped past the worn gunwales as Greg, a veteran of these waters, casually guided the skiff northward. Cook, I reflected, had no such pilot. And yet he had produced accurate charts of one of the most tortuous coasts in the world. My respect for his achievements was deepening.

We entered calmer water at cliff-rimmed Freshwater Cove, where other fishermen from Brigus South were hauling their nets. The damp cold had penetrated my clothes. Gulls and puffins squawked overhead as the men scooped silver-hued cod — some thousands of them — into the dories.

"Aye, we've always respected Captain Cook here," said Greg. "He was a bloody good seaman, and his charts saved a lot of lives over the years."

In Taunton, England, at the Royal Navy's Hydrographic Department, I examined some of those original charts: neat, precise, factual, lightly tinted, signed with a flourish "Jam^s Cook." "Cook was a superb hydrographer, a master of the craft," Admiral Haslam told me. "He paid meticulous attention to detail, and yet the charts are cohesive and attractive. Although he made charts everywhere he went, I'd say his work in Newfoundland was about his best — one of the most impressive bodies of work I've ever seen."

For Cook, the routine of summer surveying and winter reporting went on, with a few unusual events. In the summer of 1764, a large powder horn exploded in his right hand, leaving a long scar that ran as far as the wrist. He settled a minor crisis when three bored seamen got drunk. In November, he crossed the cantankerous North Atlantic in the *Grenville*, his first long voyage as a commander. In December, before Christmas, Elizabeth Cook bore another son, Nathaniel.

In 1766, the first of Cook's Newfoundland charts were published—with Admiralty permission, but at his own expense — and some would serve mariners for more than a century. Fortunately, he did not lose by his venture. Sometime in 1767, Mrs. Cook gave birth to a daughter named Elizabeth.

The most consequential event of these years came on an August day in 1766. Thick fog had blanketed the west coast of Newfoundland for weeks. It parted just in time for Cook to observe an annular eclipse of the sun, which he recorded with his usual precision. He had a good telescopic quadrant, and used it to calculate the longitude for his place of observation. Determin-

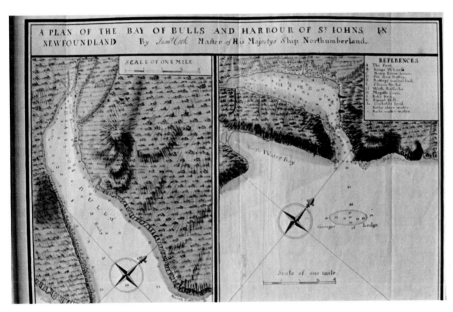

ing longitude with any approach to accuracy was the most difficult aspect of navigation in that age; and Cook applied this sophisticated method with astonishing success. That winter he submitted his observations to James Bevis, a physician interested in astronomy. Dr. Bevis belonged to the Royal Society, the single most influential body of savants in England. He presented Cook's findings before the Society on April 30, 1767, calling him "a good mathematician, and very expert in his Business. . . ."

Now the name of a master in the Royal Navy was recognized not only by the Lords of the Admiralty, but also by the Fellows of the Royal Society. Among these were men of great wealth and high station, men who chatted of botany and astronomy with the king himself. In April 1768, as James Cook planned yet another summer in Newfoundland, other people — powerful and highly placed people—were preparing plans for him. Plans that in just five months would dispatch him on a voyage to paradise.

"go to Newfoundland ... to be employed in making surveys of the Coast & Harbours of that Island...."

Honing the hydrographic skills he learned mapping the Quebec Basin, Cook drew detailed and accurate charts of Newfoundland (above). As master of a navy ship, he surveyed 2,000 miles of the island's fogbound coast in the summers from 1763-1767.

OVERLEAF: *Sunset clouds stain the glassy surface of Bonne Bay on Newfoundland's western shore. Cook's diagrams revealed the hazards such placid waters hide.*

"Aye, we've always respected Captain Cook here. He was a bloody good seaman, and his charts saved a lot of lives over the years."

$\mathcal{B}$reakers lap the grim reaches of Freshwater Cove as fishermen reset their cod nets at dawn. Clarence Hawkins mends his trap with a twine needle; straining, the Hawkins brothers haul in their catch. In a ten-hour day they might net 20,000 cod (right). The Grand Banks of the North Atlantic, cornerstone of Canadian fishing today and vital to the economy of 18th-century western Europe, prompted French-British rivalry that led to war. Cook's charts, pinpointing Newfoundland's rich waters and alerting sailors to the consistently foul weather, helped the British expand their fisheries. To the Admiralty and the Royal Society, the surveyor proved himself an explorer, and —recording an eclipse —even an astronomer. His experience, judgment, and intelligence destined Cook for greater voyages.

ASIA

North Pacific Ocean

NORTH AMERICA

North Atlantic Ocean

Plymouth

EUROPE

Equator

Society Islands

Tahiti

SOUTH AMERICA

AFRICA

AUSTRALIA

Tuamotu Archipelago

Rio de Janeiro

South Atlantic Ocean

Cape of Good Hope

0   3000
STATUTE MILES

South Pacific Ocean

Cape Horn

Tierra del Fuego

The First Voyage:
"The Entrancing Cloud-hung Heights of Tahiti"

Greatness was thrust upon the shoulders of James Cook at a time—and in a way — that he least expected it. Unbeknown to him, the Royal Society and the Admiralty had been collaborating for months on a scheme to send a naval vessel and skilled specialists "to the Southward" in order "to observe the Passage of the Planet Venus over the Disk of the Sun on the 3rd of June 1769. . . ."

This celestial event occurs at fixed intervals of 8, 121.5, 8, and 105.5 years. In 1679 the great astronomer Edmund Halley had explained how it might be utilized: Accurate observations from several points around the globe could yield a basis for reckoning the distance between earth and sun. That figure would form a good unit for measuring nothing less than the size of the universe—in Halley's words, "a problem the most noble."

"The 18th century was a marvelous time for science," historian Howard T. Fry told me. A distinguished, graying Englishman of 62 years, he teaches at the James Cook University of North Queensland, in Australia. "The pursuit of knowledge was a noble calling, and vast sums were spent on expeditions to gather diverse information from far corners of the world. It was not inconceivable to mount a long voyage—financially and spiritually backed by the Crown—simply to record an astronomical happening."

"But surely," I said, "there had to be other underlying reasons for such a grand venture."

"Oh yes," Professor Fry responded quickly. "You see, the British Empire was entering a period of tremendous upthrust. After the Seven Years War, the English hoped to expand greatly their sphere of influence. In particular, they wanted to preclude French moves into the South Pacific. Voyages of exploration would yield new lands for the spreading Empire, as well as new trading centers and military bases. Besides the usual forms of competition between nations, there was also a kind of scientific competition. Prestige would come to those individuals—and their homelands—who discovered and reported on unknown continents, peoples, plants, and animals."

Over the centuries some curious notions had developed concerning *Terra Australis Incognita*, a vast southern continent that geographers thought must exist in the Pacific Ocean. Its character owed more to fancy than to fact. Supposedly it possessed fabulous riches of gold and silver, and a people graced with charm and nobility. Marco Polo brought his own tales of such a place when he returned to Europe from the Far East in 1295. Subsequent explorers added to the legend. Mercator and Ortelius, the luminary 16th-century cartographers, actually detailed the continent on their world maps. Spanish, Portuguese, Dutch, English, and French navigators all ventured into the Pacific hoping to be the first to discover this unique land. Obviously, whichever nation first made friendly contact with the eloquent natives would have a splendid advantage.

Belief in the Unknown Southern Land continued well into the 18th century. Just three months before Cook sailed, Capt. Samuel Wallis of H.M.S. *Dolphin* returned to report his discovery of Tahiti. Offshore one evening the watch had seen "the tops of several mountains the Extreems bearing from South to S.W. upwards of twenty Leags . . . the long wishd for Southern Continent, which has been often talkd of, but neaver before seen by any Europeans." Wallis had been too ill to explore this desirable land. He would have been chagrined to learn that he was a victim—not the first and certainly not the last—of a beguiling Tahitian sunset.

Perhaps the greatest advocate of the Southern Continent theory in the

PRECEDING PAGES: *Remote in the sapphire Pacific lie islands of the Tuamotu group. Cook sailed through this area just days before sighting Tahiti, the base for astronomical work that the government announced as his party's official mission.*

34

18th century was Alexander Dalrymple, a Scots-born hydrographer and explorer. He accepted the logical argument that a continent was "wanting on the South of the Equator, to counterpoise the land on the North." Laboriously, he calculated an appropriate size for it — 3,640 square degrees. He found a location for it in an unexplored region, which placed it in the temperate zone. He concluded that there were "in the southern hemisphere, hitherto totally undiscovered, valuable and extensive countries, in that climate best adapted for the conveniency of man . . . ."

An intense and impressive young man, of broad vision and short temper, Dalrymple presented himself to the Royal Society as a suitable commander for the proposed voyage to the South Pacific. As a skilled navigator he had credentials for observing the transit, he had sailed on voyages of survey and discovery in the service of the East India Company, and he had commanded a small ship in Asian waters. The Society gave him its formal endorsement.

The Lords of the Admiralty demurred. Appointing a

Antarctic winds chill Tierra del Fuego, where an Indian family huddles in its hut of sticks and leafy branches. Cook stopped here briefly in January 1769 for fresh water, firewood, and any foodstuffs he could find. Tragedy struck during one foray ashore when two black servants, caught in a sudden snowstorm, died of exposure in the night.

"perhaps as miserable a set of People as are this day upon Earth."

civilian would be "totally repugnant to the rules of the Navy." Since the Admiralty was providing the ship, the crew, and the provisions, their Lordships wanted a naval officer. According to Dalrymple, they offered a compromise—he would control much of the expedition but leave direction of the ship to its appointed captain— and he refused it. The Admiralty held firm, and Dalrymple resigned from the venture.

"Although circumstances prevented him from commanding this great voyage," Professor Fry told me, "it certainly didn't halt him. He went on to become the first Hydrographer to the Admiralty and a great proponent of the expansion of British trade. He was truly an exceptional man."

The awkward question of command was settled almost in routine by Philip Stephens, secretary to the Lords of the Admiralty, who presented a new candidate. Weighing the influential opinion of Hugh Palliser, their Lordships, in April 1768, chose a man whose name had appeared more and more often in a context of compliments: James Cook. The Royal Society approved, and on May 5 officially appointed him an observer of the transit.

Just 20 days later came another notable distinction. No longer a mere warrant officer, he now held His Majesty's commission as a first lieutenant in the Royal Navy—a promotion that must have pleased mightily the likes of Thomas Skottowe and John Walker, Hugh Palliser and Lord Colville. Lieutenant Cook proudly took command of the ship bought for the mission, the *Endeavour*: a welcome and familiar craft, a cat-built bark from Whitby. In size and rigging, she was nearly a twin to the Walkers' *Freelove*.

The next few months must have seemed like a tempest. Cook surely spent as much time as possible with his wife and young family, but hours were taken up with outfitting the *Endeavour*, obtaining the necessary astronomical and navigational equipment, and meeting his new crew. In addition, he made final arrangements about payments for his Newfoundland charts and planned for a Yorkshire cousin to stay with his children and Elizabeth, who was far advanced in her fourth pregnancy.

On July 30, Cook left them, sailing to Plymouth to wait for a special complement—the civilians who were to accompany him as naturalists and astronomers. The term "scientist" would not acquire its modern distinction for another century, and Cook referred to these specialists as "experimental gentlemen." Preeminent among them was Joseph Banks, at 25 already a distinguished traveler and natural philosopher. Wealthy, well educated, self-confident and energetic, Banks had a wide-ranging interest in science, although his passion was botany. Later he would become President of the Royal Society, a power in London society and in the international realm of learning.

With Banks came Dr. Daniel Solander, a respectable and well-liked pupil of the brilliant Swedish botanist Carl von Linné, immortal as Linnaeus; two sensitive and talented artists, Sydney Parkinson and Alexander Buchan; H. D. Spöring, a gravely serious secretary from Sweden; four servants and two big dogs. And the Royal Society had sent astronomer Charles Green as a paid observer.

All told, 11 civilians, 71 naval personnel, and 12 marines were crowded onto the *Endeavour*, a ship just 106 feet long and 29 feet in the beam—barely wider than a singles tennis court. Her departure was delayed while carpenters provided extra cabins, understandably tiny.

Finally, on a cloudy August 25, 1768, the *Endeavour*, her sails billowing with a propitious breeze, slipped her moorings at Plymouth. All the company, Banks noted in his journal, were "in excellent health and spirits perfectly prepard (in Mind at least) to undergo with Chearfullness any fatigues or dangers that may occur in our intended Voyage."

Cook may soon have retired to his cabin to review his orders—marked "Secret"—from the Admiralty. I can visualize the youthful captain sitting alone at a roughhewn desk in the great cabin. As he took up the oilskin

Tahiti's jagged peaks rise behind Matavai Bay in this view from Point Venus, where Cook set up his observatory in May 1769 to watch the planet Venus cross the disk of the sun on June 3. Off nearby Moorea, three men paddle a racing canoe.

"the women
begin to have a
share in
our Freindship
which is
by no means
Platonick."

*C*oconut oil heightens the sheen of a Tahitian girl's skin as she
prepares for a dance. Famed since Cook's visits for beauty, grace,
and sensuality, Polynesian women charmed, tempted, and cajoled
his crewmen. One astronomer wrote later: "Virtue is held in little esteem
here, the women gladly . . . swiming to the ships sides and getting
on board for the sake of a Nail or a bit of old cloth. . . ." Because one
item of metal could win the favors of an island girl, Cook had to

keep his men from wrenching the nails out of the Endeavour's planking. But
easy romance had a tragic consequence: venereal disease. Cook heard with relief
that not his men but earlier explorers —from France —had introduced it. Still,
always concerned about the welfare of the island peoples, he feared that the disease
might "in time spread it self over all the Islands in the South Seas, to the eternal
reproach of those who first brought it. . . ." He politely refused the overtures of even
the loveliest girls; some teasingly called him "Old, and good for nothing."

pouch, he may have paused for a moment and listened to the snap of the sails as they caught the breeze, the creak of the hull as it slipped through the waves, the muted chant of the sailors at their tasks. A ripple of excitement, tempered with pride and self-satisfaction, must have welled up inside him as he unfolded the crackling parchment and read the clear script directing him to exotic and faraway lands—lands that other lads from Yorkshire had never heard of, let alone had a chance to explore.

Now he was to "make the best of your way to the Island of Madeira and there take on board such a Quantity of Wine as you can Conveniently stow, ...and proceed round Cape Horn to...King Georges Island...." King George's Island was Tahiti; Wallis had recorded its location, and the Royal Society and the Admiralty agreed that it would serve perfectly as a base.

By Opunohu Bay on Moorea, boys ride home carrying breadfruit, a staple in the South Pacific. People "seldom make a meal without some of it," as Cook noted. Below, young women of Rurutu in the Austral group weave mats of pandanus leaf, another basic resource. Islanders cut the long stiff leaves from the tree, dry them, soak them in seawater, dry them again, then cut them into long strips for braiding (detail, right).

Further "Instructions," even more secret, came sealed. After the observation, Cook was to proceed south "to make discovery of the Continent" or any other lands. He was to "take Possession for His Majesty" of uninhabited country or of any "Convenient Situations" that "Natives" would allow him. He was to chart the land; to gather specimens of minerals or gems, ores, seeds, fruits, and grains; and to cultivate "Friendship and Alliance" with the local populace. He was also to claim such "Islands . . . that have not hitherto been discover'd by any Europeans," and to survey any of importance.

If the scope of his task daunted him, Cook made no mention of it. But surely he must have considered it warily. He was embarking on a voyage of extraordinary magnitude: one fragile wooden ship to combat the unpredictable dangers of the Pacific Ocean; one small band of men to confront untold

Tahitian fishermen on Matavai Bay try for one more catch as the sun sets behind

Moorea. The outrigger's design remains little changed from that of craft described by Cook. 43

thousands of potentially dangerous natives; and one solitary man to bear the final responsibility for the mission. Doubt, worry, and even fear must have plagued James Cook at times, but he never revealed as much. He was resolute in the face of adversity, an example of strength to his men.

Rolling along at a speed that never exceeded eight knots, the *Endeavour* cruised to Madeira. Hard gales drowned several dozen fowls, aboard to supply fresh eggs, but the nanny goat that gave milk for the officers was unhurt. At Funchal harbor the deep-laden bark found room for 3,032 gallons of excellent wine, 270-odd pounds of fresh beef, and "a Live Bullock" for slaughter at sea. Cook also purchased onions "in plenty," 20 pounds per man. Then, skirting Tenerife in the Canary Islands, the *Endeavour* began her long run southward across the Atlantic in the northeast trades.

Shipboard life, varied mainly by weather, seemed tolerable. Crewmen swapped tales with new shipmates. The naturalists eagerly observed such novel phenomena as the swimming of small "blubbers" or jellyfish, Portuguese men-of-war, flying fish. The artists drew and painted new creatures. Cook and Green constantly compared their lunars—multiple sets of complex observations and calculations designed to indicate longitude. Humid weather announced the tropics—leather trunks and books turning white with mold, pocketknives and razors rusting into uselessness.

Just below the Equator Cook deferred to maritime custom for a celebration: dunking into the sea anyone crossing for the first time. This rite included everyone—"Gentlemen passengers," officers, "young Gentlemen" (midshipmen), "People" (crewmen), down to the ship's cats. Those who declined because of rank, like the captain, or personality, like some of the men, had to pay a forfeit. Cook paid in rum. Banks paid in brandy for himself, his servants, and his dogs. In late afternoon the ship hove to in calm weather; the novices were hoisted to the mainyard on a rope-and-pulley rig and then allowed to drop some 35 feet into the water; and, noted the master, "The Evening was spent merrily without Debauch."

After calling at Rio de Janeiro, the *Endeavour* coasted South America toward Cape Horn, mainly in fine summer weather. Christmas came, and with it another naval tradition. As Banks recorded it, "all hands get abominably drunk so that at night there was scarce a sober man in the ship, wind thank god very moderate or the lord knows what would have become of us." The captain, who undoubtedly kept an eye on ship and weather, merely noted that "the People were none of the Soberest."

As latitude increased, so did the cold — and the surprisingly diverse wildlife. Men delighted in spotting turtles, albatrosses, penguins, even krill. Cook set Banks and Solander ashore at Tierra del Fuego, and they returned with a collection of plants, "most of them unknown in Europe," according to Cook, "and in that alone consisted their whole Value. . . ."

Farther along this coast, at the Bay of Good Success, the *Endeavour* anchored to take on wood and water. Thirty or so Indians gathered on the beach to see Cook and his men—the first of the many primitive peoples Cook would encounter during the next decade. In his usual factual and thorough way, he described the appearance, scanty clothing, and flimsy housing of these men and women, who seemed "a very hardy race," but were "perhaps as miserable a set of People as are this day upon Earth." Here a snowstorm caught Banks and a botanizing group well inland. Midday next, some returned exhausted "and what was still worse," said Cook, "two Blacks servants to Mr Banks had perished in the night with cold. . . ."

In a change of luck, surprisingly favorable winds conveyed the *Endeavour* past Cape Horn with relative ease. Often, in adverse conditions, this passage could take months—months of struggling against frigid storms, mountainous waves, and looming icebergs. Cook made it in 33 days, exercising prudence among ill-charted fog-blurred islands in "those seas so much dreaded for hard gales of wind." Even in 1936, when Alan Villiers doubled the Cape in an iron sailing vessel with a rig stronger and more sophisticated than the *Endeavour*'s, he found it challenging: "The wind roared in the rigging, and the few sails still set picked the ship up and flung her along. Sea after snarling sea thundered at her, often throwing wild eruptions of spray and spume right over her. . . . I was 37 days on the way, and thanked God, and the shade of Captain Cook watching at my shoulder."

Cook made northwest from Cape Horn, swinging far from the South American coast and watching for signs of Terra Australis. A large continuous swell from the southwest, still running 30 hours after a gale, gave him a seaman's "proof that there is no land near in that quarter." Banks, consulting a copy of Dalrymple's map of the unknown continent, saw that they were comfortably sailing through what ought to be solid land.

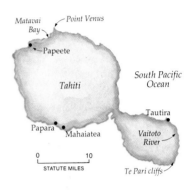

The little ship was making 95 miles a day with a fine breeze, 23 in a calm, 99 with fresh gales, 111 with a steady breeze. The weather continually improved, and the temperature rose. The men seemed generally cheerful and healthy, thanks in large part to Cook's insistence on dietary and sanitary regulations. In Cook's own boyhood such long voyages were marked by wholesale sickness and death attributed to various illnesses, but especially to scurvy.

This debilitating disease, brought on by a lack of ascorbic acid, results in fatigue, weakness, swollen spongy gums, uncontrollable bleeding, and eventually death. A ship long at sea might lose as many as two-thirds of her company to scurvy. To compensate, captains crammed in extra crewmen—which worsened their sanitary conditions.

The Scottish naval physician James Lind had published a treatise as early as 1753 identifying the causes and the cure of scurvy; but it was not widely read. James Cook kept his men free from the disease by forcing them to eat fresh fruit and vegetables wherever possible, and by providing bizarre foods thought to be antiscorbutic—as some of them actually were. He had sailed with vast quantities of these: half a ton of "portable soup," cakes of gluey meat extract to be boiled with oatmeal or pease; a malt preparation, "80 Bushells" of it; enough "Sour Krout," or fermented cabbage, to feed 70 men two pounds apiece every week for a year.

At first the seamen balked at these, but Cook applied psychology — "a Method I never once knew to fail." He ordered the sauerkraut served every day to the officers and "left it to the option of the Men either to take as much as they pleased or none atall. . . ." Within a week he had to ration it for everyone. As he said, "such are the Tempers and dispositions of Seamen in general that whatever you give them out of the Common way. . . it will not go down with them and you will hear nothing but murmurings gainest the man that first invented it; but the Moment they see their Superiors set a Value upon it, it becomes the finest stuff in the World and the inventer a damn'd honest fellow."

Cook, risen through the ranks, understood the men of the lower deck.

At every landing he tried to procure fresh foods: "scurvy grass" — greenstuff found in Tierra del Fuego and New Zealand—and wild celery and fruits wherever they could be found. Throughout his voyages he attained an incredible health record: not one death from scurvy, only a few reported

Tahiti remains today the place of legend it became after the 1760s, when explorers from Britain and France made it famous as an isle of tropical beauty and ease. Until France established her influence here in the 1840s, it figured in the long rivalry of these two nations. That struggle for power inspired the secret orders for Cook's first voyage: From Tahiti he should sail on to seek —and if possible claim for his king —the great continent thought to exist somewhere in southern waters, and record everything of value and interest he could discover.

45

cases under his eye. He was quick to act upon early signs of the disorder. He also required clean, ventilated cabins for his officers; well-washed bedding and underwear; and adequate sleep for his men.

Glorious April weather ushered Cook to his first Pacific island—Vahitahi, a pristine atoll on the eastern edge of the Tuamotu Archipelago. Coconut trees waved above glittering sands, and a group of near-naked Polynesians brandished clubs as the *Endeavour* slipped past. For a week she sailed leisurely through the dazzling isles of the Tuamotus.

Even today these islands are resplendent. A domelike sky of cobalt blue, studded with powder-puff clouds, embraces the tiny atolls. Undulating expanses of pure white sand surround lagoons of breathtaking brilliancy—blues ranging from indigo to aquamarine intermingle in a thousand subtleties of hue. Shimmering sunlight dances on the water and yields a golden vibrancy to the fronds of coconut palms.

"If the paradise of the South Pacific exists anywhere these days," Gordon reflected as we island-hopped among the Tuamotus, "it exists in tranquil lagoons like these."

But not all here is beauty. The Chevalier de Bougainville, who fought opposite Cook at Quebec and who sailed through these waters one year before him, termed the Tuamotus *l'archipel dangereux.* And dangerous it is. Reefs, shoals, and coral heads rear from the depths without warning; when the sea is calm and the sunlight oblique, it is impossible for a sailor to see such barely submerged obstacles until almost upon them.

Cook wound through the Tuamotus unscathed, and soon the entrancing cloud-hung heights of Tahiti hove into view—nearly eight full months after the *Endeavour* had left Plymouth. King George's Island—Otaheite, as Cook wrote down the native name — was truly a paradise for these sea-weary men. Under easy sail they ran on to Matavai Bay, a graceful curve of sand and palms on the north shore where Wallis had anchored before them. Rumpled hills the color of emeralds gave way to lavender reaches as mountains, mostly obscured with cloud, rose thousands of feet above the bay. The sensitive painter Sydney Parkinson considered the landscape with an artist's eye: "The land appeared as uneven as a piece of crumpled paper, being divided irregularly into hills and valleys; but a beautiful verdure covered both, even to the tops of the highest peaks."

The Tahitians were as appealing as their land. Tall, handsome, and noble of bearing, they had lustrous black hair and glistening brown bodies. The men were muscular and lithe, the women soft and lissome. They were gracious to the foreigners, accepting of their presence. Collectively, their one bad trait seemed to be a constant desire to steal. As soon as the anchor plopped down in Matavai Bay, canoeloads of people approached, some bearing coconuts and bananas. Cook noted that "it was a hard matter to keep them out of the Ship as they clime like Munkeys, but it was still harder to keep them from Stealing but every thing that came within their reach, in this they are prodiges expert."

(Continued on page 53)

"the Ships compney had... been very healthy owing in a great measure to the Sour krout, Portable Soup and Malt...."

Foods to ward off scurvy at sea form a display worthy of Cook's table: steaming "portable soup," a tub of sauerkraut, wild greens rich in vitamin C. A clean red kerchief wraps a bunch of "scurvy grass." Above, Cook's observatory at Tahiti stands re-created at Greenwich, in Britain's National Maritime Museum.

*T*ahitian dowagers attend a rehearsal for the yearly festival honoring the arrival of the first missionaries. A choir leader energetically directs his group as others wait to sing. In 1797 — just 28 years after Cook's first visit — Protestant evangelists came from England on the ship Duff; they tried to establish their own ideas of dress, to end infanticide, and to abolish such pagan customs as polygyny. Roman Catholic missionaries came in the 1830s, with French support. Today, most Tahitians profess Christianity.

OVERLEAF: *Double canoes break from the starting line for a major race on Tahiti's southern coast. Such spirited contests, highly popular, inspire a growing pride in the traditions of the Polynesian past.*

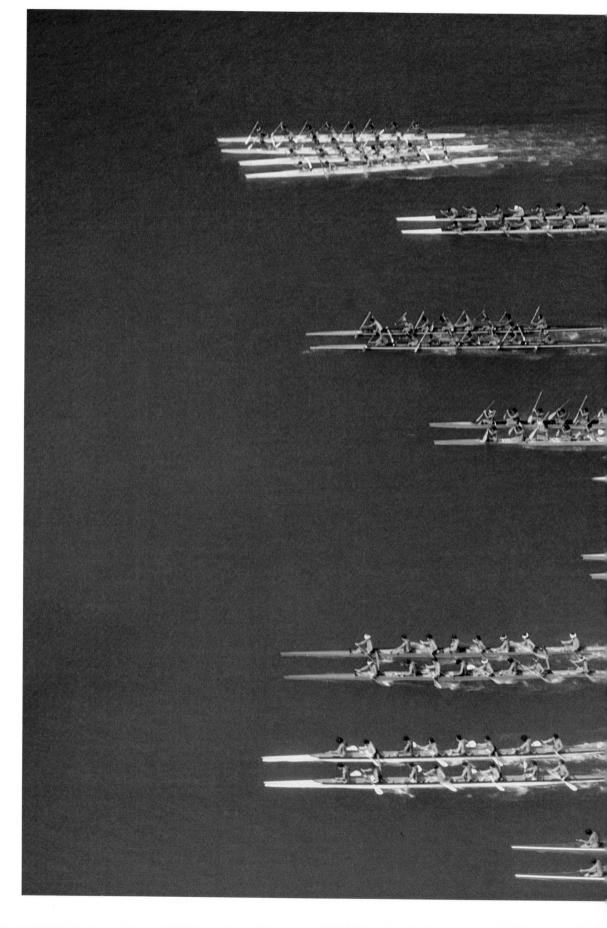

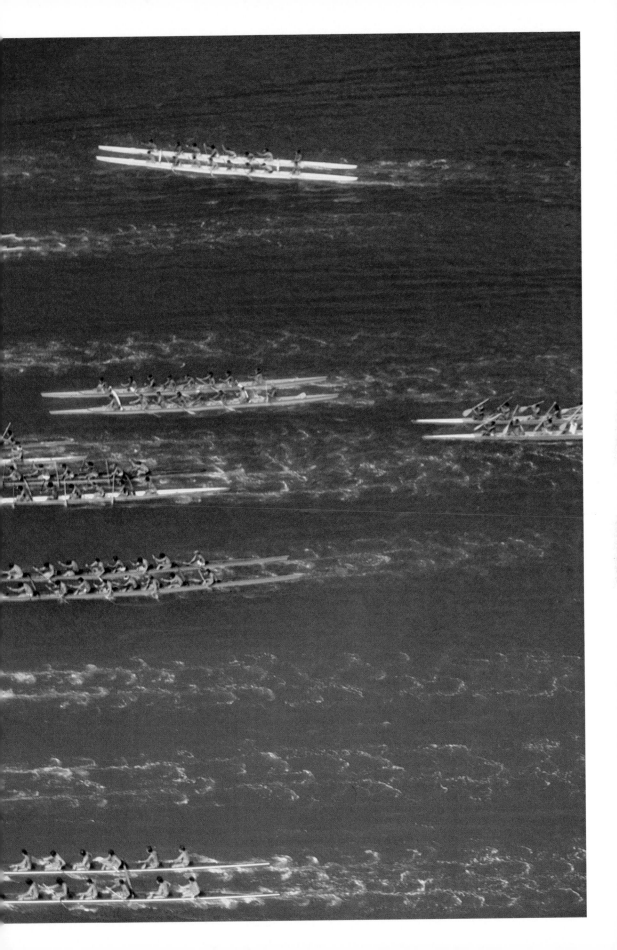

Hoping to prevent "confution and quarels," Cook circulated rules for dealing with the "Indians," as everyone called the natives. He ordered all the ship's complement to attend only to duty on shore; to respect the Tahitians always, cultivating friendship "by every fair means"; and to permit only designated individuals to trade. The final rule: "No Sort of Iron, or any thing that is made of Iron, or any sort of Cloth or other usefull or necessary articles are to be given in exchange for any thing but provisions."

A people of Stone Age technology, the Tahitians had an insatiable longing for any implement made of metal. The seamen soon learned that the gift of a single nail could earn the favor of a Polynesian enchantress for the night. And a hatchet could enliven fantasies born during an eight-month voyage. These Tahitian women were not reticent; they would clamber nimbly aboard the *Endeavour* and offer themselves with joyous candor and wanton sensuality. The ship's master, Robert Molyneux, circumspectly recorded that "the women begin to have a share in our Freindship which is by no means Platonick," and Banks wrote tenderly of a "very pretty girl with a fire in her eyes."

The *Endeavour* came better prepared for the situation than the *Dolphin,* two years before. Wallis's crewmen surreptitiously stripped the *Dolphin* of so many nails, spikes, and bolts that her very structure was gravely endangered. Cook had brought extra nails, fishhooks, and mirrors; even so, he needed stern measures to protect them.

Almost immediately, he set his men to preparing an observation post for the transit of Venus. On the northeast end of Matavai Bay a long palm-shaded peninsula juts into the water; here, on Point Venus, he decided to establish his headquarters. Within two weeks, the men had constructed Fort Venus—complete with embankments, a palisade, and emplacements for eight small cannon.

Cook chose wisely: Point Venus is an ideal spot to observe anything. One warm March evening I sprawled on the long charcoal-gray strand and watched the sunset. Soft waves rhythmically purred on the beach, chuckling and hissing quietly as they rushed toward me and then receded. Spindly coconut palms and wispy casuarinas played in the delicate breeze. From far away came the melodic sounds of a guitar and a gentle Polynesian song. From farther came the baying of a dog. The sun slowly nestled behind the craggy heights of Moorea, an island just northwest of Tahiti. Its peaks split the sunlight into bands that splashed Matavai Bay with fiery orange and mauve. Just as the light began to fade, a lone boatman in a dugout canoe paddled across the bay. Despite the multitude of changes that have overtaken Tahiti in the last two hundred years, at that unforgettable moment I sensed the heartbeat of the ancient Polynesia.

With the establishment of Fort Venus, Cook thought himself "perfectly secure from any thing these people could attempt." Of course, he wasn't.

"I eat taro because I love it. It is my coffee after I have worked hard all day."

Leon Tepa of Rurutu works his crop of wetland taro. Its tubers supply carbohydrates and proteins; its young leaves and stalks, provitamin A and vitamin C.

OVERLEAF: *Bora Bora from the air reveals the classic reef where ferocious surf forced Cook to veer away in July 1769, and the single channel closed to him by adverse winds. Today an airstrip built in the lagoon allows easier access to the island.*

Light-fingered pilferers made off with anything portable, including Cook's socks, which he was using one night as a pillow. They filched Bank's matched pocket pistols and powder horn, his white jacket and waistcoat trimmed with silver braid. They took a spy glass from Dr. Solander. They picked a snuffbox from the surgeon's pocket. They even stole the large and cumbersome astronomical quadrant from under the nose of a sentry. This was calamity. The quadrant, a superb precision device made of brass and bell metal by the famed craftsman John Bird, was essential for the observations. It had to be retrieved. This brought Cook, for the first time, into direct conflict with the Tahitian chiefs.

Already the English had identified the man who seemed highest-ranking on the island, a chief or *ari'i* whose name they heard as Tutaha or

Joseph Banks, this wealthy and influential naturalist, zealously collected specimens on the first voyage. He hired artists Sydney Parkinson and Alexander Buchan to record such novelties as the tropical fishes at right. Though he had a falling out with Cook, he acquired many objects from Cook's later voyages and from other explorers; famous men of learning met at his home in London to study these exotic prizes.

Tuteha. With Tupaia, a distinguished priest and adviser, Tuteha was staying close to Point Venus—which seemed to seal the friendship between the two groups. Now Cook's first thought was to seize Tuteha as a hostage for the safe return of the quadrant; he suspended the order at news that Banks was on its trail, and set off after Banks with a small guard. In the woods near sunset he met Banks returning with the quadrant; but back at the fort they found that Cook's order had been misunderstood and Tuteha detained by force. Cook freed him immediately, but Tuteha was shamed and angry. Courteous supplication and gifts — broadcloth garments, an ax — finally soothed him. Such encounters, however, left Cook confused by the structure of Tahitian society and the rules by which it functioned.

"*C'est très compliquée*, very complicated, this ancient Tahitian society," mused historian Paul-R. Moortgat. We lunched on traditional Tahitian dishes—raw fish marinated in lime juice and served in coconut milk, chicken cooked with taro leaves—in a restaurant overlooking the harbor of Papeete, busy capital of the sweep of islands now grouped as French Polynesia. "Tahiti had developed a social and political system complex enough to have evolved toward aristocracy," he explained. "The traditional society was fundamentally inegalitarian. No single authority governed the country, and Captain Wallis at the time of his stay counted 17 'districts.'"

This situation led to some of Cook's misunderstandings. Arriving from a Europe of monarchies, he expected to meet one accepted sovereign; instead, he came across several powerful individuals, some of whom seemed to outrank others. Indeed, marriage—and war—between districts resulted in constantly shifting alliances and rulers. When Wallis discovered Tahiti in 1767, a woman named Purea seemed to be "queen." When Cook arrived, her power had waned, and Tuteha—apparently insignificant before—was a paramount chief.

"Here life and all its activities were marked by religion," M. Moortgat continued. "Religion was all-powerful; and combined with the weight of the social order and a lack of contact with other peoples, it did not create conditions favorable for an evolution of society."

Undoubtedly the confusion that beset Cook and his men had its counterpart among the Tahitians. Surely they puzzled over the antics of the Englishmen who built an odd "fort," lugged shiny heavy equipment about, and picked all sorts of plants—only to put them into transparent "jars."

Friendly transactions prospered, however. Trade in metal objects, cloth, and various beads and trinkets brought the English the bounty of a tropical isle: small hogs, plantains, coconuts, bananas, breadfruit, fish, taro, yams, sugarcane, and various fruits. Commerce encouraged efforts at learning a strange new language. Tahitian friends attended divine service with Banks one Sunday, standing or sitting or kneeling when he did. Tahitian musicians, on hearing an English song, gave it "much applause."

But still there were those striking differences. Cook's party were shocked to discover that many children born to parents of high rank were suffocated at birth. And the forthrightness of young women appalled—but also enticed—the more priggish Englishmen, who were slower to notice the reserve shown by maidens of high status. Once some attractive girls with male attendants approached Joseph Banks. A man gave him fruit; another arranged pieces of bark cloth on the ground. A young woman "stepd upon them . . . unveiling all her charms . . . [and] turning herself gradualy round. . . ." She repeated this twice before the cloths were offered to Banks as a gift. Politely, he offered presents in return.

HALICHOERES RADIATUS

Such gifts of cloth figured in a ritual called *taurua;* but this event seems to show simply that the girls admired Banks, a handsome and engaging young man who indeed had found himself a "flame," as he said, in Tahiti.

ABUDEFDUF COELESTINUS

Finally, after seven weeks of preparation, the day of the transit, June 3, drew near. Spöring had improvised some missing parts of the quadrant, using watchmaking tools brought from England, and restored its delicate balance as well as he could. Cook had it mounted on a large cask, filled with "wet heavy sand" for stability. Clocks were set up nearby. On the 2nd, Molyneux recorded that "the Captain and Mr Green is entirely employ'd getting every thing compleatly ready. . . . every thing very quiet & all Hands anxious for Tomorrow." Cook worried about the weather. But the "day prov'd as favourable . . . as we could wish," wrote the captain; "not a Clowd was to be seen the whole day and the Air was perfectly clear. . . ."

PARACIRRHITES FORSTERI

HENIOCHUS CHRYSOSTOMUS

For six long hours a black spot edged across the disk of the sun. The only disturbances came from heat blasting to 119°F., and "an Atmosphere or dusky shade round the body of the Planet." That atmosphere made it almost impossible to tell the instants when the planet's image was tangent to the sun's rim, and timing those precisely was all-important. Still, Cook and his party had done their best. The observations, professionally recorded, could be collated in Europe— a process which, unfortunately, proved inconclusive.

KYPHOSUS INCISOR

Thus the acknowledged aspect of Cook's mission was complete. Now for the adventuresome work under the sealed orders. First, however, he had things to accomplish: preparing the ship and charting the island. The ship's bottom was "boot-topped"—scraped clean, then coated with pitch and brimstone. Sails and rigging were mended, spars varnished, supplies checked. That work had proceeded slowly, routine broken by a few notable events. The artist Buchan had died of an epileptic fit. Banks, with the unshakable self-assurance of a wealthy gentleman, blackened himself with charcoal and bedecked himself with a scrap of bark cloth and took part in a religious ceremony. Sydney Parkinson asked the natives to *tattow*—tattoo—

Fishes of tropical waters, their colors recorded fresh-caught by Buchan (top) and Parkinson, appear above identified by the scientific names accepted as valid today.

Parkinson's floral sketches show his deft hand —stilled by death on the homeward trip in 1771. From a port in the Dutch East Indies to the Cape of Good Hope, malaria or dysentery took him and 29 shipmates.

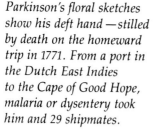

HEBE STRICTA

THESPESIA POPULNEA

EUGENIA SUBORBICULARIS

Coronilla coronata Willd.
Aeschynomene speciosa.
striata

Sydney Parkinson pinx. 1767.

SESBANIA GRANDIFLORA

his arms in the Tahitian fashion, a gesture worthy of a sailor. The officers and gentlemen dined on Tahitian dog; they agreed, wrote Cook, that "they Never eat sweeter meat, we therefore resolved for the future not to despise Dog's flesh."

Cook fretted over the spread of a "Venerial distemper" among the natives and his own men. A month before his landfall he had ordered the surgeon to conduct an inspection for possible quarantine. The Tahitians implicated Bougainville's crew, relieving Cook's sense of responsibility but not his concern. Till his death, as far as humanly possible, he tried to safeguard the peoples he encountered.

On the morning of June 26, James Cook stepped into the *Endeavour's* pinnace, a small sailboat, and with Banks and a Tahitian pilot set out "to

make the Circuit of the Island in order to examine and draw a Sketch of the Coast and Harbours thereof." On foot where the terrain permitted, in the pinnace the rest of the way, they explored the land, making wild stabs at spelling its place-names. In places, the shoreline is gentle; in others, cliffs plunge into the sea in a torment of rock and water — and nowhere more wildly than at the Te Pari cliffs of the Tautira district, isolated at the far eastern end of the island.

I visited Tutaha Salmon, the mayor of this district, in his spacious home by the harbor called Mouillage de Cook — Cook's Anchorage. I mentioned that the most powerful person the captain met on his first stay was also called Tutaha, and the mayor smiled. An intense man with a pencil-thin mustache, he told me of Tautira. "Even today we are more traditional than the rest of Tahiti. We are farther from Papeete and the government; we run things the way we, the people, want. Therefore it is quieter, more peaceful, more Tahitian."

Captain Cook, a keen observer, gathered artifacts that rank as treasure today. A Polynesian mourning dress of bark cloth has feather tassels and a feather cloak. Pearl shells brighten the crescent-shaped wood chest piece; slips of mother-of-pearl adorn the upper apron. Feathers of tropic birds edge the headpiece mounted over the face mask of shell. The wood carving above, which Cook gave his great official patron Lord Sandwich, probably decorated the stern of an Austral Islands canoe.

I remarked that Cook had observed many dugout canoes in this region. "Ah, the pirogues," Tutaha answered quickly, "they are still important to us. We in Tautira build our canoes in the old way — the way our grandfathers taught us. That is why we always win the races." Tautira canoe teams are famous throughout the Pacific for their competitiveness. Invariably Tutaha's pirogue teams have won the right to represent Tahiti in the annual races in Hawaii — and they have consistently carried off a prize.

"Ah, it's because we follow the old techniques more than our other

OVERLEAF: *A fisherman off Tubuai in the Austral group — one of the areas where the islanders still rely on sail — runs with a breeze of the southwest trades.*

Polynesian brothers. We build our canoes by hand right here and we teach our boys the techniques from the old days. We are good because we have learned from the past." And good they are—Tutaha's office is filled with first-place trophies from scores of races.

With his men rowing the pinnace, Cook skirted the rugged Te Pari coast, which he found "broken and danger[ou]s and had it not been for our pilot we should have found it difficult to proceed...." Also with a local pilot, Gordon and I sailed around the Te Pari cliffs at sunrise. An amethyst glow produced by roiling black clouds hung among steep valleys, even though the water sparkled with dawning sunlight. The glancing angle of the sun created a vibrant rainbow in the valley of the Vaitoto River, framed by gray sawtooth ridges and fronted by white sand and greenish-gold palms. As the rainbow intensified in color, a pair of small white birds winged gently past and disappeared—and then the rainbow too was gone, a magical instance of the evanescent beauty of Tahiti.

Circling to the south of the island, Cook came to Papara, where he discovered "a wonderfull peice of Indian Architecture"—the grand *marae* of Mahaiatea. On a court 377 feet long, paved with great flagstones, a huge stepped altar rose to a height of some 50 feet. The shrine honored "queen" Purea's son Teri'irere as high chief of Papara. Cook had met this dignitary one day when a man came to Fort Venus carrying a lad of seven upon his back. All the Tahitians in the vicinity "went out to meet them, having first uncover'd their heads and bodies as low as their waists." Recognizing this "as a ceremonial Respect," Cook thought it was directed to the man. Told that the boy was the object of this homage, he concluded that this must be "Heir apparent to the Sovereignty of the Island." In reality he was but one ari'i among the others.

Two days of meandering among reefs and shoals brought Cook back to Matavai Bay. He had finished the hundred-mile journey in a week, producing a chart that "I believe is without any material error."

Now it was time to depart. The sailing season for higher latitudes was at hand. Fort Venus was struck down for firewood, supplies were stored, final repair work was completed on the *Endeavour.*

Then two marines, Clement Webb and Samuel Gibson, disappeared. Cook heard "that they were gone to the Mountains & that they had got each of them a Wife & would not return...." He could not permit flagrant desertion, and knew he would need Tahitian help to find his men. He took hostages—a few persons of rank, including Tuteha—to be held safely aboard the ship. Within a day the love-blind marines were back.

Yet damage had been done to carefully cultivated friendships. The hostages came ashore looking "sulky and affronted," as Banks noticed. "Thus we are likly to leave these people in disgust with our behavour towards them, owing wholy to the folly of two of our own people," lamented Cook. The captain, Banks, and Dr. Solander—all popular among the Tahitians—paid a courtly visit to Tuteha the night before they sailed. According to Banks, they enjoyed "a perfect reconciliation." Cook's estimate was cautious: Apparently the Tahitians understood "our friendly disposission . . . several of them were in tears at our comeing away."

Tears undoubtedly were shed on July 13, when the *Endeavour* weighed anchor and sailed out of Matavai Bay. Cook and his men had spent three months in Tahiti, and each had a feeling of his own for the land and the people. Tahiti, always, would own something of each man's heart.

Under a cloud tinged with saffron, a three-masted square-rigger built to the style of Cook's adventurous lifetime rests in calm seas at day's end off the coast of Tahiti.

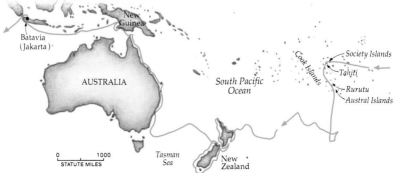

The First Voyage:
"Into Uncharted Seas"

A shimmery blue sea nestling undiscovered realms beckoned Lt. James Cook and the *Endeavour* to the southward. Before chasing phantom continents, however, Cook sailed northwest to explore islands of which the Tahitians spoke favorably. Now he had a pilot.

Aboard the *Endeavour* was the high priest Tupaia, who wanted to join the voyage and visit "Pretane"—Britain. The captain, who respected Tupaia's expert knowlege of the islands and thought him "a very intelligent person," had refused to take him. The Admiralty had not authorized anything like this, and might not accept the responsibility. Banks could afford it, and did: "Thank heaven I have a sufficiency and I do not know why I may not keep him as a curiosity, as well as some of my neighbors do lions and tygers ... the amusement I shall have in his future conversation and the benefit he will be of to this ship ... will I think fully repay me."

Tupaia, who belonged to a family of famous navigators, soon proved his worth. After sending down a diver to tell him the ship's draught, he helped guide her safely through reef-strewn shallows. He produced favoring winds through prayer to his god Tane, although Banks was positive that he prayed only when he sensed an approaching breeze. And he served as an excellent mediator and interpreter at the islands that soon appeared.

The Society Isles Cook named them, "as they lay contiguous to one a nother." Huahine, Raiatea, Tahaa, Bora Bora, Maupiti—as gleaming a strand of pearls as ever adorned an ocean. For three weeks Cook cruised among them, making more Polynesian friends, renewing supplies of food, and claiming territory for the Crown.

Soon, though, the *Endeavour* and her crew were ready to venture into uncharted seas—as Banks put it, "in search of what chance and Tupia might direct us to." After four days of sailing due south, a small, high, coral-fringed island rose on the horizon, a solitary sentinel in a trackless ocean. Rurutu belongs to the Austral Islands, thinly scattered bits of French Polynesia that dot the Pacific south of Tahiti. Tupaia knew of the island, but here his presence proved of little help; the natives tried to seize the pinnace as it neared shore, and only musket fire could dissuade them.

Today the people of Rurutu are friendly and engaging, and their island is one of the most traditional in all of French Polynesia. Most people speak only their melodious dialect of Tahitian, and the old ways still flourish. Nearly every day, women gather to weave mats, rugs, and hats from dried strips of pandanus leaves. Flashing fingers manipulate long beige strips into swirling designs that soon result in exquisite round mats or hats of a dozen different shapes. Babies lie in mothers' laps, and little girls tentatively assist their elders. Nonstop chatter punctuated by peals of laughter resounds through the small structures where the women work.

Tradition permeates all facets of life in Rurutu, especially agriculture. Early one morning, Gordon and I met young Leon Tepa as he farmed his taro patch in a boggy swale near the coast. With its starchy tuber, taro in its many varieties has long sustained Polynesians as a staple food. The so-called "wet taro" flourishes in marshy earth, and we squished through acres of rich mud to Leon's patch.

"I grow taro only for my family," said Leon, a lean, shy man of 26 years. Wielding a stick, he pried a foot-long taro from the mud; with one stroke of a machete, he separated the tuber from its long stalks and heart-shaped leaves. These he carefully stuck back in the mud. "An ancient Polynesian custom," he said. "When you pick one taro, you must plant another." The earth is so fertile and the climate so benign that the stalks promptly put out

PRECEDING PAGES: *Day's end at Dusky Sound on New Zealand's South Island evokes the evening on which Cook discovered —and named —it: March 14, 1770.*

roots, and within a year a new taro is mature. In 15 minutes Leon had gathered a dozen taro, enough to feed his family for several days. They boil it and eat it with fish, or pound it into thick, flavorful poi. Gordon asked Leon if he ever grew tired of taro. "Oh no," Leon exclaimed, "I eat taro because I love it. It is my coffee after I have worked hard all day." Captain Cook, who sampled taro throughout the Pacific, found it "a very wholsome root."

After circling Rurutu, Cook sailed on south in search of the continent. Tupaia told of other islands nearby, but as Cook recorded, "we can not find that he either knows or ever heard of a Continent or large track of land." Early on August 16, Cook "saw the appearances of high land to the Eastward bore up to wards it, but at 10 we discover'd it to be only Clowds. . . ."

For two more weeks the *Endeavour* voyaged south to latitude 40°—some 1,500 miles from Tahiti—without noticing even the illusion of land. Here, Cook's orders directed him to turn westward and make for "the Eastern side of the Land discover'd by Tasman and now called New Zeland."

Abel Janszoon Tasman, a competent Dutch explorer, had coasted part of the western shoreline of New Zealand in 1642 and 1643 during an epochal voyage that also added parts of Australia, Tonga, and Fiji to the sketchy map of the world. Those were great years for the mariners and merchants of the Netherlands, and Tasman's orders stressed gain rather than geography. He recorded the coast he saw in fair detail, but only as a squiggly line in unexplored ocean.

Cook accomplished much more. In a running survey lasting six months, he mapped and thoroughly explored the 2,400 miles of New Zealand's perilous shoreline. He coped with the howling winds of the roaring forties, the tricky currents of the Tasman Sea. He had to prowl coasts never visited by Europeans, where a rock, a shoal, or a reef might be lurking anywhere below the surface. He had to enter harbors where a shift of wind might pen him up for days. He had to weigh the risk of being "embayed" in such a harbor against the risk of foul weather on open water. He had to battle storms, and he had to endure flat calms when even a gentle tide could carry his ship against a cliff. He had to keep the *Endeavour* in repair, his men healthy and in good spirits, and any menacing natives at a safe distance.

Incredibly, he did it all — almost perfectly. Although slips and discrepancies appear in his charts, they are all forgivable. As Admiral Haslam told me, "Cook's survey of New Zealand is one of the most awesome accomplishments in the history of cartography and hydrography—especially when you consider the difficulties he faced and the rudimentary equipment that he had to work with. It's truly astonishing."

Following Admiralty orders to explore "as great an Extent . . . as you can," Cook surveyed New Zealand's coastline of more than 2,400 miles in six months. From Poverty Bay he ran briefly southward, then turned north to circle the North Island. Via Cook Strait he returned to sight Cape Turnagain; he swung south and rounded the South Island before heading west toward Australia.

"Stretching" — that is, sailing continually and easily under "crowd of canvas"—Cook made westerly toward New Zealand. During this time, the gentlemen updated journals, cataloged specimens, and completed sketches, maps, and paintings. The captain's great cabin, which Cook graciously shared with the others, must have been jammed to the ceiling with men and their collections—it measured, roughly, 18 feet by 12. It must also have provided a forum for lively discussions on many subjects — botany, anthropology, geography. "Now do I wish," wrote Banks, "that our freinds in England could by the assistance of some magical spying glass take a peep at our situation: Dr Solander setts at the Cabbin table describing, myself at my Bureau Journalizing, between us hangs a large bunch of sea weed, upon

$\mathcal{M}$orning mist fills the vales near Mercury Bay, where the Endeavour's specialists watched the planet Mercury cross the sun on November 9, 1769. The first phase of the transit, wrote Cook, "was Observed by Mr Green only. I . . . was

taking the Suns Altitude in order to ascertain the time." On hills and headlands in this region the Maori inhabitants had built villages fortified for war; Cook inspected one such *pa* with keen interest and thought its defenses showed "great judgement."

the table lays the wood and barnacles. . . ." It was an intellectually stimulating environment for many of this ship's company.

For none more than for Cook. Since boyhood he had been learning the "practical part" of things. Now he was living among men thrilled by learning for its own sake. A farm lad's knowledge of nature would help him follow the eager discussions of philosophers trying to fit each new plant, each new sea creature, into the grand order of creation. A chart-maker's skill would help him assess the artists' sketches. He and Banks exchanged journals. He copied some of Banks's entries, like a good apprentice imitating his master; and thereafter he usually worked out his own accounts for himself. In short, James Cook was discovering new worlds of the mind.

Ornately tattooed, this Maori tribesman portrayed by Sydney Parkinson shows a warrior's poise; a man of today enacts the grimace of a dance to intimidate the foe. The hei-tiki carefully drawn above—an ornament carved from the treasured greenstone called pounamu—resembles the one opposite, presented by Cook to George III; eyes made of haliotis shell gleam like those of the dancer.

Others found stimulation in other ways. As often as they could, seamen would tap the liquor casks and get staggering drunk. Cook's journals record punishments for drunkenness, and the occasional death caused by drink. Boatswain's mate John Reading "was found to be very much in Liquor" one night; this was nothing unusual, so he was put "to Bed where this morning about 8 oClock he was found speechless and past recovery." He died "At 10 AM."

Possibly the men were driven to drink by the state of their food. Despite Cook's stress on cleanliness, vermin of unspeakable sorts invaded the bread. "I have often seen hundreds nay thousands shaken out of a single bisket," Joseph Banks said. "We in the Cabbin have however an easy remedy for this by baking it in an oven, not too hot, which makes them all walk off, but this cannot be allowd to the private people who must find the taste of these animals very disagreable, as they every one taste as strong as mustard or rather spirits of hartshorn [ammonia]."

By October 1769 signs of land were unmistakable: sea weed, a seal, more and more birds. About 2 p.m. of the 7th, by ship's time, a boy named Nicholas Young spotted land on the horizon, land that Banks thought "the Continent" at last. Although Cook must have suspected it was New Zealand, he could not be positive until he saw land that Tasman had described, in late December. Meanwhile, he said nothing. Master's mate Richard Pickersgill entitled several of his early charts "Part of the So Continent."

Gentle breezes on the 9th conveyed the *Endeavour* into a spacious bay, rimmed by white cliffs and green hills. Cook saw "Indians" on shore and paddling about in canoes; he hoped to establish the same congenial relations he had enjoyed with the Tahitians. His hopes were soon dashed.

Although they derive from Polynesian stock, the Maori of New Zealand did not share the Tahitians' easy ways. They were a warlike, aggressive, and defiant people.

Hoping to trade for fresh food and water, Cook went ashore with several crewmen and began walking toward a group of huts. Suddenly, four Maoris jumped from the woods and tried to seize his boat, a yawl with four boys in it. A seaman in a second boat, offshore, fired a musket over the raiders' heads; they stopped, puzzled, then resumed their approach. They ignored a second shot. Just as one man jabbed his spear at the yawl, a third shot struck him dead. At this, Cook wrote, "the other three stood motionless for a minute or two . . . wondering no doubt what it was that had thus killed their commorade. . . ." They dragged the corpse away, then abandoned it, and Cook and his men returned to the *Endeavour*.

After this inauspicious start, Cook landed next morning with armed marines and with Tupaia, who addressed the natives in Tahitian—"it was an [a]greeable surprise to us to find that they perfectly understood him." They made their point with action vividly described by Lt. John Gore: "with a Regular Jump from Left to Right and the Reverse, They brandish'd Their Weapons, distort'd their Mouths, Lolling out their Tongues and Turn'd up the Whites of their Eyes Accompanied with a strong hoarse song, Calculated in my opinion to Chear Each Other and Intimidate their Enemies. . . ."

Tupaia warned the Englishmen to stay alert as 20 or 30 warriors came up around them. The Maoris snatched at the strange weapons, taking a light sword; shots were fired and four men fell wounded, one mortally.

Another encounter resulted in more Maori deaths. Cook was disconsolate. "I am aware that most humane men who have not experienced things of this nature will cencure my conduct . . . but . . . I was not to stand still and suffer either my self or those that were with me to be knocked on the head."

To avoid further casualties, Cook sadly left this bay, "named *Poverty Bay* because it afforded us no one thing we wanted." In honor of the boy who had first sighted New Zealand, Cook named a rumpled peninsula to the south Young Nick's Head; the lad also won a gallon of rum.

The *Endeavour*'s bad fortune continued. She ran southwest in search of water and food, to no avail. Finally, with "the face of the Country Vissibly altering for the worse," the captain had done enough. At Cape Turnagain he reversed course, and luck began to change. The natives became peaceable, Cook traded cloth for food, and about 50 miles north of Poverty Bay he found a fine harbor.

A strong Pacific surge still crashes on the rockbound southern shore of Tolaga Bay, where Cook anchored near a grassy cove with a clear stream. The crewmen set to work filling casks and cutting firewood. The naturalists, spurred by the boyish enthusiasm of Banks, eagerly began botanizing — their first real opportunity in this new land — and Cook, of course, charted the bay and explored inland. Parkinson noted that this lovely area "with proper cultivation, might be rendered a kind of second Paradise."

"They . . . distort'd their Mouths, Lolling out their Tongues and Turn'd up the Whites of their Eyes . . . ."

All that the Maori produced from this fertile earth, it seemed, were taro and "sweet Potatous and Yamms," but they gladly traded their rootstuffs and fish for tapa from Tahiti, while their priests discussed theology with Tupaia. The Englishmen found the Maori men indifferent to iron items and their women modest — though Banks called the ladies "as great coquetts as any Europæans could be and the young ones as skittish as unbroke fillies." Cook, as usual concerned with diet, plied his men with "Sellery and Scurvy grass."

A week's coastwise sailing northward brought him to a large harbor where he established a base for observing the transit of the planet Mercury across the sun on November 9. If weather allowed the observation, he could work out the longitude of "this place and Country." The transit was duly recorded by Green and Cook, and Cook noted happily that the "situation of few parts of the world are better determined than these Islands are. . . ."

Along these coasts, Cook considered with special interest the fortifications constructed by the warlike Maori. Perched on a steep bluff overlooking Mercury Bay was such a *pa*, or fortified village. Protected by deep ditches and high palisades, it could enclose dozens of people during a brief attack or a long siege. From its wooden parapets the defenders could hurl spears or stones. The veteran of the Louisbourg-Quebec campaign analyzed it carefully—no spring inside, but a stream nearby; stocks of dried fish and fern roots within. A "very strong and well choose post," he thought; here "a small number of resolute men might defend them selves a long time against a vast superior force. . . ." From the ship he had seen many larger towns of this sort, and concluded very justly that "this people must have long and frequent wars, and must have been long accustom'd to it. . . ."

Before dawn one mild April morning, Gordon and I climbed a lofty hill behind Mercury Bay. Where Cook had found a rough landscape and a warrior people, I saw only pastoral gentleness. The rising sun fired a low-lying mist and touched the grassy downs below us with gold. Horses frolicked and sheep grazed in meadows abutting small, neat farms. To me, it evoked the Yorkshire of Cook's childhood.

"Sharp pick, sharp shells, no gloves," says oysterman Mike Stewart, prying barnacles off a clump of the Auckland rock oysters found only in tidal waters of northern North Island. Gloves would impede the handwork that yields him 350 dozen oysters on a good day. Along the coasts near Mercury Bay, shellfish in abundance provided many a feast for Cook and his crew.

In addition to other bounties, Mercury Bay gave his crew "an immence quantity of Oysters." I've sampled oysters elsewhere, but never had as delectable a feast as I did one morning with Mike Stewart, one of the few hand-pick commercial oystermen left in New Zealand. In his chugging work boat, we motored into the Firth of Thames on the west coast of the Coromandel Peninsula.

I asked Mike about oysters, and he smiled. "They are like women, incomprehensible to most men," he said. "And each an individual and must be treated as such." A lean man roughened by years of work at sea, Mike has crisp blue eyes and a thick beard showing a touch of gray. Over the years he has continued to learn—"and I do not consider that I know it all yet. I love eating oysters," he went on, "but I get tired of opening them. If I could find someone to open them for me, I'd eat them forever."

Along the shore of Double Island, he set to work chipping oysters from rocks exposed at low tide. "To get them up, some you hit hard, some soft." He cracked a couple open, washed them in the sea, and handed them to Gordon and me. We could only agree with Mike—they had a succulent briny taste, and I could eat them forever.

"the Long-boat ...return'd ...loaded as deep as she could swim with oysters...."

Like many New Zealanders today, Mike holds James Cook in the highest esteem. "Yes, my bloody oath, he was probably the greatest sailor who ever lived," Mike said as he bent to his work. "He wasn't the first, but he was the most." I often wonder how the austere Yorkshireman would react to such testimonials as these.

Two centuries ago he made a lasting impression. In Mercury Bay, a Maori boy named Te Horeta boarded the *Endeavour* with his elders. Years later he recalled: "There was one supreme man in that ship. We knew that he was the lord of the whole by his perfect gentlemanly and noble demeanour. . . . He was a very good man, and came to us—the children—and patted our cheeks, and gently touched our heads." Cook gave Te Horeta a nail, a possession he treasured for years afterward. The boy and his companions thought of the Maori proverb: "a noble man—a *rangatira*—cannot be lost in a crowd."

As the Polynesians knew well, any man can be lost at sea. For the next four weeks, Cook battled winds of hurricane force, and prodigious waves and currents, as he rounded the northern point of New Zealand. He identified Tasman's Three Kings Islands and Cape Maria van Diemen, and finally ascertained that this indeed was New Zealand.

Luckily the weather relented for Christmas Day. Banks jauntily wrote that "Our Goose pye was eat with great approbation and in the Evening all hands were as Drunk as our forefathers usd to be upon the like occasion." It turned squally again as the ship inched down the west coast to the "very broad and deep Bay" indicated as Zee Haens bocht on Tasman's chart. On the ragged southern coast of this bay Cook found an ideal harbor, "a very snug Cove" where he would anchor for three weeks.

Ship Cove he called it, a nick in the land along a sinuous finger of water. It had neither the sublime elegance nor the sumptuous bounty of Matavai Bay, but it offered abundant wood and "excellent" water, fish for the netting, strand for careening a ship, and a climate reminiscent of England's at its best. Cook would return to it repeatedly on later voyages.

*H*ere the land was "one intire forest," and it is today. Driftwood litters the small pebbly beach cut by a burbling stream. I dipped a hand into these cool waters — although I had not been living on long-stored water from a wooden cask, I could imagine how Cook savored this crisp clean drink. Steep, wooded hills rising behind a grassy meadow seem to embrace the cove. It is a tranquil spot, with a charm Banks discovered after just two nights there: "This morn I was awakd by the singing of the birds ashore .... their voices were certainly the most melodious wild musick I have ever heard, almost imitating small bells but with the most tuneable silver sound imaginable...." New Zealand bellbirds — graceful, green, and small — keep to the shelter of green foliage where they lighten the spirit with their tinkling song.

West of Cape Brett, luminous hills of Motuarohia Island extend into the Bay of Islands cove now named for Cook; here he and his men stepped ashore in November 1769. When attacked by Maori warriors — here and elsewhere — he restrained his own crew and succeeded in restoring friendly dealings. Today, a southern black-backed gull glides over tranquil waters at evening.

The Maori of Ship Cove greeted the *Endeavour* with a hail of stones. But Tupaia talked with them, and they proved friendly enough although it became clear that these people were cannibals. On many occasions, Cook reported, Tupaia tried to argue the Maori out of this custom, "but they always as strenuously supported it and never would own that it was wrong." Visiting a nearby cove, Cook saw a group preparing their provisions and recognized human forearm bones cleaned of flesh "which they told us they had eat." Convinced by a plethora of grisly evidence, Cook named the spot Cannibal Cove.

One afternoon he made a troublesome climb to a hilltop for a vantage of the wooded countryside — and the true nature of New Zealand lay revealed. Tasman's "bay" was actually a strait that split the land in two. Although Cook was too modest a man to dwell on this circumstance, it was probably Banks who insisted on naming this strait for its discoverer.

On January 31, Cook and a party rowed over to an isle called Motuara. Cook gave gifts to the people, and Tupaia told them the strangers wanted to set up a mark of their visit. The seamen erected a post on the highest ground and "hoisted thereon the Union flag and I dignified this Inlet with the name of *Queen Charlottes Sound* and took formal posession of it and the adjacent lands in the name and for the use of his Majesty, we then drank Her Majestys hilth in a Bottle of wine...."

Some 210 years later, Gordon and I scrambled to that summit. From the

PRECEDING PAGES: *Off Cape Brett, this rock "with a hole perced quite thro'" moved Cook to make a pun — on the name of Sir Piercy Brett, a Lord of the Admiralty.*

In the snow-bright panorama of the Southern Alps, New Zealand honors heroes of discovery. Mount Cook National Park contains 22 peaks higher than 10,000 feet; grandest of all, Mount Cook itself rises to 12,349. Visibly the highest in the aerial view below, it justifies its Maori name: Aorangi, the Cloud-Piercer. A closer approach reveals its jagged wind-whipped summit (below, left). Although Cook often glimpsed rugged snow-covered heights as he sailed along the South Island's

"mountains piled on mountains to an amazing height".

western coast in March 1770, foul weather apparently kept him from seeing these imposing peaks. Second highest among them at 11,475 feet, Mount Tasman and its namesake glacier commemorate Abel Janszoon Tasman, the Dutch seaman who in 1642 became the first European to reach New Zealand. He explored the western coasts, primarily along the North Island. Daunted by the Maori, none of his party went ashore; but Tasman made useful sketches, and a map, of the coasts he saw.

depths of his camera bag, Gordon produced a bottle. He uncorked it with a flourish, raised it in salute, and offered a toast to the memory of Captain James Cook. We drank that toast with gusto in fine New Zealand claret.

Thereafter the *Endeavour* cruised the east coast of the South Island, chasing the continent that was never more than cloud. On March 10 she rounded South Cape with a favoring wind — "the total demolition," said Banks, "of our aerial fabrick calld continent."

Instead, the explorers turned their attention to the reality of the west coast, and they found it formidably beautiful. This fiordland region is one of the most wildly romantic in the world. Long gray fingers of the sea probe among precipitous rock walls; brooding clouds hang among the heights. Adverse winds prevented Cook from investigating the fiords, but one seen at evening, Dusky Sound, lodged in his memory. Soon the *Endeavour* was passing one of the world's grand mountain ranges — the Southern Alps. Thick weather, with fog and drizzle, often blurred the view, but sometimes white summits shone above the clouds. Cook rightly suspected that the mountain chain extended from one end of the island to the other. He wrote of a "ridge of Mountains which are of a prodigious height ... cover'd in many places with large patches of snow which perhaps have laid there sence the creation." The crown of this lofty and jagged range soars to 12,349 feet. The Maori call it Aorangi, the Cloud-Piercer; an English surveyor named it Mount Cook, and it is as stern and rawboned as its namesake.

By the end of March the *Endeavour* was near Cook Strait again, completing the circumnavigation of New Zealand. James Cook was ready for new discoveries. Earlier, pressed by the naturalists to linger in some appealing harbors, he had refused. He was unwilling to risk embayment, and loss of time, for a few new species: "I had other and more greater objects in view, viz. the discovery of the whole Eastern Coast of New Holland."

*T*hat a sprawling, generally barren expanse of land existed in the southwest Pacific had been known to geographers since the early 1600s. Dutch explorers had mapped bits and stretches of the north, west, and south coasts—but of the east nothing was known. Cook was determined to fill in the blanks.

The *Endeavour* crossed the Tasman Sea without incident and within three weeks closed the southeastern coast of Australia near Point Hicks, or today's Cape Everard. She turned north, and Banks noted apparently bare places in the hilly landscape: "it resembled in my imagination the back of a lean Cow, coverd in general with long hair, but nevertheless where her scraggy hip bones have stuck out farther than they ought accidental rubbs and knocks have intirely bard them of their share of covering."

Soon the ship took a narrow entrance to a spacious bay. Cook named it Botany for "the great quantity of New Plants &cᵃ Mʳ Banks & Dʳ Solander collected in this place." Although the southern suburbs of Sydney encompass the bay now, many of those plants still flourish here.

"There has been a tremendous influx of foreign plants," botany teacher Fred Stubenrauch told me. "But I'm always astonished at the variety of native plants still to be found. Many of them are hardy and tenacious, and they've resisted the onslaught of the new." Early one morning, I joined Fred and a group of his students from Cronulla High School on a field trip to Botany Bay, the young teacher as energetic as his pupils.

"Probably the predominant plants that Cook and Banks found were the eucalypts, the famous gum trees of Australia," Fred explained. "And an important genus of trees and shrubs is *Banksia,* in honor of Sir Joseph." In the soft morning light the students scattered to search for native plants. Their trophies included a sprig of banksia; common bracken, small fern;

dusky coral pea, a low creeper with delicate red flowers; and a flax lily, with a purplish flower and leaves that the local people wove into baskets.

At the British Museum (Natural History) in London, Gordon and I had the pleasure of seeing first-voyage material from the Banks Collection with an acknowledged authority on the subject, the respected botanist William T. Stearn. Formerly a senior principal scientific officer with the museum's botany department, he has worked with these riches for decades. He brought out folio volumes of watercolors by Parkinson so we could compare likenesses rendered from living plants with specimens dried for preservation two centuries ago. These, in remarkable condition, can be compared to detailed descriptions penned by Solander.

"Scientifically, the *Endeavour* was the best-equipped ship ever to set out from Europe up to that time," said Dr. Stearn, a vigorous man with a brushy mustache. "And it returned with the most extensive collections of botanical material ever seen—thousands of items and no fewer than 1,300 new species. Much of the credit can be attributed to Banks's unbounded enthusiasm; but much credit also must go to Dr. Solander, who had a highly specialized but broadly based knowledge. Between the two of them—and with Cook's interest and assistance—they made a great contribution. This collection stood as the standard of reference for almost a century."

I asked Dr. Stearn to assess the influence of the *Endeavour*'s voyage.

"A tremendous impact, and particularly on science. For instance, it began and established the Admiralty's policy of including naturalists on long voyages of discovery. The ultimate result, of course, was the extraordinary work of Charles Darwin on the *Beagle*."

While the *Endeavour*'s party worked at Botany Bay, with Banks carefully drying his specimens "upon a sail" spread in the sunlight, Cook saw a few Aboriginals. He found that "neither us nor Tupia could understand one word they said." They had crude small canoes, and seemed to subsist on shellfish grubbed from the mud. They fled with unmatchable speed from their strange visitors. "I think them a timorous and inoffensive race," Cook concluded. All "they seem'd to want was for us to be gone."

Charting Australia's east coast, checked only by the Great Barrier Reef's 1,250-mile coral maze, Cook sighted Point Hicks on April 19, 1770, and departed through Torres Strait four arduous months later.

Cook was gone, after displaying the English colors and carving an inscription on a tree. In generally fair weather, the *Endeavour* wound north, Cook charting the coast and naming headlands and bays. Forays ashore introduced the explorers to some of the plagues of this coast—biting ants, vicious mosquitoes, sand burrs, and stinging green caterpillars that waited, said Banks, like a "wrathfull militia" among the mangroves. The sea, too, was becoming dangerous. Shoals, sandbanks, and islets were constant hazards; and the ship cleared some underwater obstacles near Cape Capricorn with only two feet to spare. Cook began sending his boats ahead to take soundings.

Unwittingly, the ship was groping her way into a trap—the Great Barrier Reef, one of the most astonishing of nature's marvels and one of the world's most treacherous waterways. For hundreds of miles as it meanders north to Cape Melville, the Reef pinches inexorably closer to the mainland. It creates

OVERLEAF: *Eastern grey kangaroos gather on open grassland. Such creatures bore, said Cook, "no sort of resemblance to any European Animal I ever saw."*

a tangled maze through which tides seethe and plunge. Into this menacing labyrinth Cook innocently sailed.

It was some days before he understood his peril. In the Whitsunday Passage the waters deepened, and his ship cruised easily among steep islets with sparkling beaches. Farther north, the naturalists thought they saw coconut palms. In fact, they were near the latitude of Tahiti.

Then the ship edged past a small hummocky point; in retrospect, Cook named it Cape Tribulation, "because here begun all our troubles." He could see rocks and isles ahead. Evening brought a fine breeze, and he ordered a course offshore, to sail by moonlight. The man heaving the lead for soundings reported 20 and 21 fathoms, 120 and 126 feet: more than a safe margin for a ship drawing 14 feet. The gentlemen ate supper and went to bed.

Suddenly the man with the lead had 17 fathoms, and before he could heave his line again "the Ship Struck and stuck fast." The ragged coral claws of the Great Barrier Reef had grasped the *Endeavour.*

Cook was on deck immediately — reportedly "in his drawers." He found his ship lodged solidly in coral shallows at full high tide. No efforts could budge her; the waves grated her against the reef. He gave orders to lighten ship, hoping the next high tide would lift her free. Twelve hours later came high water — but the *Endeavour* remained stranded. By noon, Lt. Zachary Hicks recorded, they "hove overboard 20 Tons of Iron & Stone Ballast & Six Carrige Guns." As the tide receded she canted to starboard and began leaking, badly. Three pumps were manned, shifts changing every 15 minutes, normal profanity hushed. The tide rose; the leak was gaining on the pumps. The ship righted herself as the water rose—and floated free.

Now she drifted in deep water, at risk of sinking there with all her gear and stranding her survivors forever. But the coral that had pierced her hull had broken off and now helped plug the leak. The crew got her under sail. Limping, her boats taking soundings ahead, the *Endeavour* edged along in search of a harbor.

On a crystalline September day, Gordon and I donned scuba gear and went diving on Endeavour Reef, scene of the mishap. Coral looms from the depths in a multitude of shapes and patterns, forming great billowing parapets through which fishes play. Long branching prongs of coral reach toward the surface in a reef some four miles in length. The sun shimmers through the azure water in liquid patches of gold, illuminating great spotted coral trout, tiny fearful wrasses, and giant clams. Underwater, the reef seemed more like a world of magic than the snare it proved for Cook.

Intensely alert, a kangaroo holds her young joey in her pouch and watches for danger. A marsupial's pouch proved to Banks the relationship to the opossum of the New World. Cook, who liked to collect birds' eggs as a boy, must have admired Australia's many vivid parrots, such as the rainbow lorikeet above. Its range extends from Tasmania north along the eastern coast.

On June 16 the *Endeavour* reached the mouth of a small river, and she staggered in. Here she would lie for seven weeks while the carpenter and his helpers worked on her battered timbers.

As the work progressed, Banks and Lieutenant Gore, an experienced hunter, penetrated the bush, hoping to bag a strange animal that some crewmen had spotted fleetingly. The two glimpsed one of the creatures, but it vaulted away outdistancing even Banks's greyhound. A week later Gore succeeded in shooting one of these beasts; then he got one a good deal larger; and the greyhound finally caught a very small juvenile. In modern terms the first was probably an eastern wallaroo; the closest Banks could get to a

Students from Cronulla High School, on a science field trip, trek through the lowlands near Botany Bay. This area gave Cook's party a rich botanical harvest in just eight days. The genus Banksia

"The great quantity of New Plants... collected in this place occasioned my giveing it the name of Botany Bay."

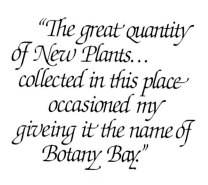

honors the naturalist who discovered it; at left, a student climbs one of these small trees to pick the sprig below. With good field technique, Parkinson drew a branch of Banksia serrata, *painted the exact greens of one leaf, the subtle grayish pink of one flower, the reddish brown of one fruit, and left the rest for completion at leisure.*

native name for all these animals was *kangooroo*. With descriptions, drawings, and specimens in hand, the *Endeavour* party would make known to science Australia's distinctive, astonishing, and endearing mammals.

To see a kangaroo bounding along the crest of a dew-covered hill at dawn is to see the grace nature bestows on its favored creatures. The roo moves with the muscular ease of a cheetah and the carefree agility of a porpoise. It has the perky face of a llama, the inscrutable look of a camel, and the curiosity of a dog. To me, it is one of the world's most winsome animals.

Near the Endeavour River, as Cook named his refuge, the gentlemen sighted more Aboriginals—as evasive as those at Botany Bay. Finally, after about three weeks, a few men were lured to the ship with gifts, including fresh fish. Cook and Banks noted their height, about 5 feet 6 inches, their slender build, their dark skin painted white and red, their close-cropped hair "of the same consistence with our hair." These men kept their women away; but the gentlemen saw one with telescopes, and Cook was shocked to see her as naked as the men, "as naked as ever she was born."

Venturing on board the ship, several men spotted some large sea turtles caught for food. They asked for one "by signs," were refused, and tried to take it anyway. Being "disapointed in this," Cook said, "they grew a little troublesome and were for throwing every thing over board they could lay their hands upon. . . ." They rushed ashore, and had a grass fire raging in an instant. Luckily, the crew got it under control, little was damaged, and the ship, though leaky, was ready to sail.

Yet the company's predicament deeply worried Cook. He feared they were caught with no way out except retracing their course, beating against the prevailing wind: "an endless peice of work." Nevertheless, he was determined to go on northward if he possibly could.

On August 4, 1770, he ventured once again into the tangle of the Great Barrier Reef. Behind, however, he left a legacy. On the final sinuous curve of the river nestles a community called Cooktown, linked to the rest of Australia by a winding, rutted road of 212 miles. "The mangrove swamps, the mudbanks—most everything's the same it was when Cook was here," town administrator Graham Gallop told me. He pointed to a column honoring the captain. "About the only changes are the town and this monument here. Our little town was turned upside down in 1970 when Queen Elizabeth II came to dedicate it."

All the citizens—some 700 of them—are fiercely proud of Cook and their handhold on the beginnings of national history. "The way me mates and me got it figured," said one boisterous Aussie in a pub, "because Captain Cook stopped here so long, this was the first real settlement in all Australia. That means that every other place in the whole bloody country is a suburb of Cooktown." Cheers and laughter echoed through the pub.

Easing out of the Endeavour River, Cook was soon surrounded by shoals with winding channels between them, all "dangerous to the highest degree." Led by the pinnace, he slowly made northward, anchoring at night. Lizard Island's 1,179-foot summit gave him a fair vantage point, and for the first time in weeks something like a welcome surprise. He could see that off to the northwest "the Sea broke very high. This . . . gave me great hopes that they were the outermost shoals. . . ." *(Continued on page 98)*

Botanist William T. Stearn of the British Museum (Natural History) displays treasures of the Banks Collection: Dried specimens of New Zealand flax (right foreground) overlie the watercolor of this same plant. Above: Parkinson's study of a lily from Botany Bay, Blandfordia nobilis, completed from his field sketch and color notes by James Miller in England in 1775.

"I named...Cape Tribulation
because here begun all our troubles."

*T*ide-washed boulders on the beach at Cape Tribulation hint at counterparts in the shallows and hidden dangers beyond. Offshore, the Great Barrier Reef's mazes of jagged coral inspired Cook to give the headland its present eloquent name.

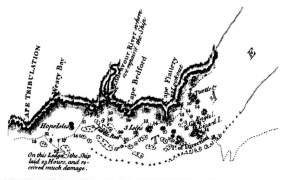

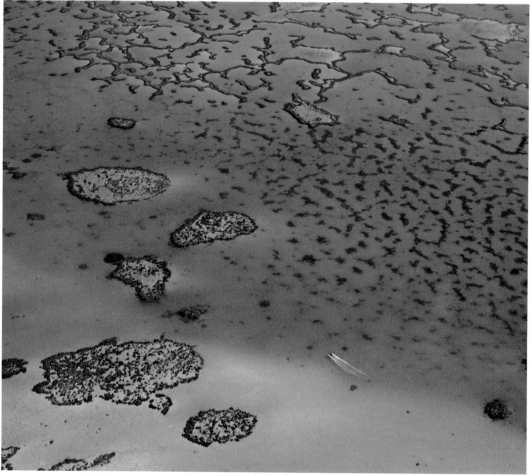

$\mathcal{P}$rongs of coral rise near the hull of a small vessel at Endeavour Reef, where Cook's sturdy bark, venturing along by moonlight, found herself stranded and leaking on an unsuspected ledge in waters that seemed deep enough for safety. Prepared from Cook's own manuscript chart, the engraved map above indicates the innumerable hazards north of Cape Tribulation; among the minor reefs Cook spaced out in capital letters a grim label: THE LABYRINTH. Seen from the air, coral crests stand out vividly in the translucent shallows typical of this coast; but the lookout on a small ship could easily fail to recognize them in deeper channels with cloud shadows on the surface. After 23 unnerving hours on the reef, lighter by many tons of jettisoned load, the Endeavour floated free at high tide—with the hole in her hull partly plugged by the very coral that had pierced it.

As crewmen set out to find a passage through the reef, the Endeavour *lies careened for repairs by her namesake river in this engraving based on an eyewitness painting. Only after she had lain there a full three weeks did any Aboriginals risk making contact with the explorers. Parkinson portrayed them well armed, as at right; Cook noted their total, unvarying nudity. Today, their world and lifeways have altered—but the* Endeavour River *still meanders gently seaward among the hills and bays of Australia's coast.*

97

The *Endeavour* inched along to that outer reef, found a gap in it, and broke into the open sea. In one of his great understatements, the captain recorded that everyone felt "quite easy at being free'd from fears of Shoals &cᵃ —after having been intangled among them . . . sence the 26th of May," nearly three endless months.

With a steady fresh gale on "a well growen Sea," the ship leaked more but that seemed a "trifeling" matter by now. Then, on the night of the 16th, "it fell quite Calm." The *Endeavour* drifted in the dark. About 4 a.m. men could clearly hear "the roaring of the Surf." At dawn "the vast foaming breakers were too plainly to be seen not a Mile from us" and the waves were taking the ship inexorably —across water too deep for anchoring —toward the wall of the outermost reef. There the ship would be splintered in seconds. Cook ordered his three small boats lowered; men toiled at the oars to tow 360-odd tons against the power of the Pacific swell. Discipline never faltered: "in this truly terrible situation not one man ceased to do his utmost," and that calmly. Mr. Green stood on deck taking observations while Cook gauged the distance to the coral: "between us and distruction was only a dismal Vally the breadth of one wave. . . ."

Then the faintest of light airs reached the sails. The ship responded, moving seaward. Another unnerving calm, and then "our friendly breeze." By noon, with the help of ebb tide and oarsmen, the ship was more than a mile from the reef but still in grave peril. Scouting in one of the boats, Lieutenant Hicks found a passable gap in the coral and the *Endeavour* rushed through it on a flood tide like a millrace. This was "the narrowest Escape we ever had," wrote Pickersgill, "and had it not been for the immeadate help of Providence we must Inavatably have Perishd. . . ." That risky passage through the reef went onto the chart as Providential Channel.

"It is but a few days ago that I rejoiced at having got without the Reef," Cook wrote that night, "but that joy was nothing when Compared to what I now felt at being safe at an Anchor within it, such is the Visissitudes attending this kind of Service. . . ."

Weary, strained by months of the most challenging conditions, and yearning no doubt for home and family, Cook set down this verdict: "Was it not for the Pleasure which Naturly results to a man from his being the first discoverer even was it nothing more than Sand or Shoals this kind of Service would be insupportable especially in far distant parts like this, Short of Provisions & almost every other necessary."

For another three days the *Endeavour* struggled among shoals and reefs. On his chart of this coast from the Endeavour River northward, Cook spaced out in capital letters the words "THE LABYRINTH." On August 21 he reached the northeastern tip of Australia, Cape York, in safety.

On a nearby island the captain "hoisted English Coulers and in the Name of His Majesty . . . took posession of the whole Eastern Coast . . . by the Name of *New South Wales*. . . ." His party fired three volleys of small arms, "Answerd by the like number from the Ship."

Confident now of a mission fairly carried out, Cook planned his course for England. Banks observed that in spite of general good health almost everyone was suffering from a disease named "Nostalgia." After a visit to the New Guinea coast, contrary winds bedeviled the passage to the teeming Dutch port of Batavia — today's Jakarta — on the island of Java. There he meant to buy supplies and to refit the *Endeavour* for her voyage home.

A low, swampy, prosperous but festering city, Batavia harbored dozens of diseases, none more deadly than malaria. The mosquito-borne pestilence struck down every member of the ship's company except sailmaker John Ravenhill, who was drunk most of the time. Fascinated at first by the sights of the city, Tupaia died regretting that he had left his homeland. Cook reported six other deaths at Batavia, and his ship "in a far worse condition than we expected." Her main keel was "wounded" in many places, planks were damaged by coral and riddled by shipworms, her crew had been sailing her "happy in being ignorant of the continual danger we were in."

His ship patched up and his crew depleted, Cook weighed anchor after more than two months. And now a severe dysentery — "the bloody flux," "the gripes"—swept through the malaria-weakened company, and the log of the eleven-week journey to Cape Town records the deaths: the astronomer Green, the sensitive artist Parkinson, Banks's secretary Spöring, the drunken sailmaker Ravenhill, "Alex^r Simpson a very good Seaman." One day after hours of squalls Cook made an ominous note: "hardly well men enough to tend the Sails and look after the Sick. . . ." In all, 23 men died after agonizing hours of cramps and stabbing pain. Ironically, James Cook's men were struck down by the diseases of the land, not those of the sea.

The *Endeavour* rounded the Cape of Good Hope and struggled into Cape Town's beautiful Table Bay with 29 men still desperately ill. The Dutch governor was benevolent, and provided hospital care, housing ashore, and supplies at a fair price. All but three of the sick recovered. A month later, the ship embarked on the final leg of her astonishing journey, but death still haunted her decks. Ship's master Robert Molyneux died, and Cook named Pickersgill to succeed him. Lieutenant Hicks, a capable and respected man, succumbed to consumption; he had brought it out from England, as Cook noted, but "held out tollerable well untill we got to Batavia." To the vacant slot Cook appointed Charles Clerke, a good-humored youth who had left the Essex countryside to go to sea.

Through the long June days the *Endeavour,* at her best a dull sailer, plodded steadily through the Atlantic waves toward home. Now and then she sighted a sail, or met another vessel and heard fragments of news. A Rhode Island whaler reported that "Old England" and her American colonies had settled their disputes over taxes, and "all was peace in Europe." Cook also learned that his letters sent from the Cape by a Dutch ship had not reached England— London gamblers were betting that he had perished at sea. He noted that "our Rigging and Sails are now so bad that some thing or another is giving way every day."

At noon on July 10, 1771, from high on a masthead, Nick Young, first to spot land at New Zealand, sighted the English coast about Land's End. A hearty cheer must have engulfed the ship, but nobody bothered to make notes of it. With a fresh southwest gale, "we Run briskly up Channell." On July 12, nearly three full years after departing, Cook "Anchor'd in the Downs, & soon after I landed in order to repair to London."

There, ever dutiful, he presented his charts, plans, drawings, and journals to Philip Stephens for submission to the Lords of the Admiralty. He referred to these papers with a painful modesty: "I flatter my self that the Latter will be found sufficient to convey a Tolerable knowledge of the places they are intended to illustrate, & that the discoveries we have made, tho' not great, will Apologize for the length of the Voyage."

Two weeks went by before he received any acknowledgment. Perhaps their Lordships had been dumbstruck with admiration.

Their arms and faces bright with the red ocher and white pipeclay used by their ancestors, Aboriginals of Queensland's Endeavour River country demonstrate spear-throwing for a yearly celebration of Cook's epoch-making arrival on this shore in June 1770.

Youthful Kokoimudji tribesmen and a silver-bearded elder gather around a riverside fire at evening, preparing for their commemoration of Cook's visit. This enlivens a festival supported by the Hope Vale Lutheran Mission, which seeks to serve more than 500 Aboriginals of the vicinity. Cook himself met about a dozen local men, all wary of the first Europeans in their experience. "The men . . . constantly carry their arms in their hands," Banks

noted, "3 or 4 lances in one" and a spear-throwing device in the other. Along the
eastern coast, he concluded, customs seemed much the same, though men of the
north built better canoes. "Tools among them we saw almost none. . . . They get fire
. . . with two peices of stick very readily and nimbly. . . ." Unlike all the other
"Indians" the Endeavour party had encountered, the Aboriginals showed scant
interest in metal articles, gifts of cloth, or trinkets. Except for items of food, "they
seem'd to set no value upon any thing we gave them. . . ." Cook called their voices
"soft and tunable," but the two sets of strangers could share little but gestures:
Nowhere else did he find a culture so totally different from his own.

AFRICA
Indian Ocean

New Hebrides
(Vanuatu)

Tahiti

Marquesas
Islands

SOUTH
AMERICA

AUSTRALIA

Tuamotu
Archipelago

Easter
Island

Cook
Islands

New
Caledonia

Tonga
Islands

Cape Town

South Pacific
Ocean

South Atlantic
Ocean

0   2000
STATUTE MILES

New
Zealand

Antarctic Circle

The Second Voyage:
"From Icebound Seas to Sun-blessed Islands"

The *Endeavour's* homecoming ignited a blaze of excitement in London. Newspapers wrote of "Mr. Banks and Dr. Solander's Voyage." Social invitations and scientific engagements showered upon the two naturalists. They were received at Court; they visited the King in his country retreat, where he examined their collections. Mr. Banks gave some rare plants to the Dowager Princess of Wales—and created a scandal by jilting his faithful fiancée, Miss Harriet Blosset. Overnight, Joseph Banks and Daniel Solander soared to the pinnacle of London fashion and notoriety.

James Cook quietly retired to his modest home in Mile End Old Town. But while society was clamoring for Banks, the official world was praising Cook. From the Admiralty he learned that "their Lordships extremely well approve of the whole of your proceedings. . . ." John Montagu, fourth Earl of Sandwich and now First Lord of the Admiralty, himself escorted Cook to the Court of St. James's on August 14 and presented him to His Majesty. Lieutenant Cook told of his voyage and no doubt explained his charts. The King, beyond a professional interest in his new possessions, had every reason to appreciate Cook and his work. In private a man of plain tastes, George III enjoyed using a telescope, had a superb collection of instruments for astronomy and navigation, worked hard to master practical details, and greatly respected courage. He was pleased to confer a promotion approved by the Admiralty, and Cook left the palace promised the rank of commander.

He must have been elated. His career held dazzling promise now. But not all was happiness. He had come home to learn that his infant son, born just after his departure, had died almost immediately. And four-year-old Elizabeth had perished only three months before his return. Eight-year-old James and six-year-old Nathaniel were hale and sturdy, however, and Cook's reunion with them and their mother must have been joyous. The man who dealt so gently with the Maori boys at Mercury Bay clearly longed for his own children, to whom he returned as an all-important stranger.

To his old friend and patron John Walker in Whitby, Cook could express a hard-earned satisfaction. He wrote to him on August 17: "the Voyage has fully Answered the expectation of my Superiors I had the Honour of a hours Conference with the King the other day who was pleased to express his Approbation of my Conduct in Terms that were extremely pleasing to me. . . ." From a summary of the voyage Cook turned to the future: "Another Voyage is thought of, with two Ships which if it takes place I beleive the command will be confer'd upon me."

The design and scope of this second monumental expedition were Cook's alone. After long consideration of the problematical southern continent, he had concluded that no "such thing exists unless in a high latitude." However, he continued, "I think it would be a great pitty that this thing . . . should not now be wholy clear'd up. . . ." He proposed to sail around the globe deep in southern waters, and to crisscross the Pacific often enough, to prove or disprove the continent's existence once and for all.

After brief deliberation, the Lords of the Admiralty approved Cook's proposals. Just ten weeks after the *Endeavour's* return, he was instructed to find and outfit ships for another expedition to the Pacific.

His near-tragedies on the Great Barrier Reef had convinced Cook that he needed two ships for safety. His first choice was the *Endeavour*, but this staunch explorer, repaired and refitted, already had been sent to the South Atlantic on supply duty. Cook inspected other cat-built barks from Whitby, and the Navy Board bought two. The *Resolution* at 462 tons was fractionally larger than the *Endeavour*, the *Adventure* at 340 fractionally smaller.

PRECEDING PAGES: *As if awaiting Cook's return, Easter Island's statues brood at dawn. His second voyage dispelled the myth of a great southern continent.*

As winter turned into spring in 1772, only one snag developed—and that from Joseph Banks. Lord Sandwich had invited him to serve again as naturalist, and Banks had promptly accepted. Unfortunately, the acclaim he had received had gotten the better of him. As Professor Beaglehole dryly observed, young Mr. Banks "had come by an unusually swelled head." He visualized himself as director of the voyage with Cook as a sort of executive officer. He gathered an entourage of 16 scientists, artists, draftsmen, and servants — including two horn players. He purchased overwhelming amounts of bulky scientific gear, and trade goods. And he demanded alterations to the *Resolution* to accommodate his retinue and his equipment. At Banks's insistence, another upper deck was added and a raised poop deck or "round house" on top of that. Sea trials proved the ship crank—unstable and

topheavy. Even in the quiet ripples of the Thames she almost capsized. Lt. Charles Clerke declared, "By God I'll go to Sea in a Grog Tub, if desir'd, or in the Resolution as soon as you please; but must say I think her by far the most unsafe Ship I ever saw or heard of."

Nearing the Antarctic Circle, the Adventure *and the* Resolution *take on ice for drinking water —"the most expeditious way," wrote Cook, "of Watering I ever met with."*

Slowly, under minimum canvas, the *Resolution* wobbled downriver to be restored to her original state. Banks was furious. When he visited the shipyard and saw the outcome, said an eyewitness, he "swore and stamp'd upon the Warfe, like a Mad Man; and instantly order'd his servants, and all his things out of the Ship." Joseph Banks would never return to the South Pacific.

Cook had returned to Yorkshire months before, in the Christmas season, to visit his family at Great Ayton and John Walker in Whitby. Mary Prowd, who had provided young Cook with candle and table so many years before, was overcome with joy. She threw her arms about him and exclaimed, "Oh honey James! How glad I is to see thee!" If embarrassed, the reserved Cook probably turned the conversation to old friends, or to the splendors of Court and the marvels of the Pacific.

By early summer, both his new ships were ready. Tobias Furneaux, a

stalwart but unimaginative officer, assumed command of the *Adventure*; as a second lieutenant, he had sailed with Wallis to Tahiti. Prussian-born Johann Reinhold Forster and his son George had been appointed as naturalists. The father was dour, pretentious, and humorless, the son bright and likeable. Astronomers William Wales and William Bayly joined the voyage; Wales had observed the transit of Venus from Hudson Bay, Bayly from North Cape in Norway. At the last moment, the Admiralty designated young William Hodges as artist. Talented and hardworking, primarily a landscapist, he had an eye for the "Romantick" — the bold and dramatic in nature. He would develop a mastery in the rendering of light.

On June 21, Commander James Cook took leave of his family—less than a year since he had returned to them —and took the *Resolution* to Plymouth where the *Adventure* was waiting. At 6 a.m. on July 13, with a fresh breeze from the northwest, the two ships left Plymouth Sound, and soon England slipped from view. More than three years would pass before Cook would sight that coast again—three years during which he would meet a myriad of challenges, from icebound seas to sun-blessed islands. During that time he would execute the most monumental voyage of Pacific discovery and exploration ever made, before or since. When he returned, the South Pacific would no longer be a vague realm of fictitious continents and islands; it would be a defined and well-mapped quadrant of the globe.

Bird in the hand of naturalist Johann Reinhold Forster falls to the skilled eye of his son George, the second voyage's draftsman. Engravings of such works, published in books about the expedition, added greatly to popular knowledge in an age of scientific curiosity. The British Museum (Natural History) houses more than 500 of George Forster's sketches, among them the Little Blue Penguin he painted in New Zealand.

Again Cook stopped at Madeira for wine and fresh food, including a thousand bunches of onions. He heard tales of a botanist named Burnett, age thirtyish, who had been waiting there to join Banks's party. Burnett vanished after hearing that Banks was not on board. "Every part of Mr Burnetts behaviour," wrote Cook in a wry summary of the affair, "and every action tended to prove that he was a Woman." Apparently Banks, in his grandiose plans, had even arranged female companionship for himself.

Womanless, the two ships arrived at Cape Town on October 30. Clerke noted that all were "in perfect Health and spirits, owing I believe in a great measure to the strict attention of Captain Cook to their cleanliness...." Cook's first objective lay in the far South Atlantic, a mysterious projection of land that the French explorer Lozier Bouvet had spotted in 1739. Through a dense icy fog, Bouvet had glimpsed a desolate, rocky headland cut by glaciers; he had decided it belonged to a greater landmass and named it Cape Circumcision. It so appears on at least two maps of the period, along with notes on sightings of ice.

Cold soon became oppressive as the two ships made south from Cape Town. On 24 November Cook issued each man heavy trousers and thick woolen "Fearnought" jackets; later he issued red baize caps. Heavy gales brought rain, then sleet, then snow. On December 10, the ships "Passed very near to a large Island of ice, which we mistook for land, at first," wrote Wales. Soon icebergs were all around, some nearly two miles in circuit and 200 feet high. Penguins, albatrosses, and petrels dotted the ice. The prospect of discovery raised some excitement. Richard Pickersgill, now third lieutenant on the *Resolution*, recorded, "We being Now across M. Bouvets track to y^e Eastw^d of Cape Circumcision, expect to find land hourerly, tho' sailing here is render'd very Dangerous...." Dangerous it was, and the ships were soon "stoped by an immence field of Ice to which we could see no end...."

"the whole Atmosphere seemed in a state of strange purterbation...."

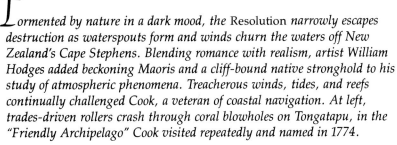

*T*ormented by nature in a dark mood, the Resolution *narrowly escapes destruction as waterspouts form and winds churn the waters off New Zealand's Cape Stephens. Blending romance with realism, artist William Hodges added beckoning Maoris and a cliff-bound native stronghold to his study of atmospheric phenomena. Treacherous winds, tides, and reefs continually challenged Cook, a veteran of coastal navigation. At left, trades-driven rollers crash through coral blowholes on Tongatapu, in the "Friendly Archipelago" Cook visited repeatedly and named in 1774.*

Hour by hour for three weeks more the lookouts stared among looming icebergs, across threatening pack ice. One morning, Cook wrote, "4 inches thick of Snow fell on the Decks the Thermometer most of the time five degrees below the Freezing point so that our Rigging and sails were all decorated with Icikles." He did not spell out what that implied. Captain Alan Villiers, from firsthand experience, has explained: "To touch the frozen rigging on frigid days was to risk frostburn which seared like flame; to fight those iron-hard sails aloft meant bloody hands and minced fingers, nails torn out by the roots and the hot blood swiftly frozen.... In the intense cold with the frigid breath of frozen hell backswept from the huge icebergs all around them, the human mechanism found places to freeze not hitherto thought of—the moisture in the eyes and in the nostrils, even among the hairs of the mustache and beard."

Swish and swirl of gaily dyed pandanus strips brighten a schoolchildren's rehearsal of a dance based on the Fijian meke. *Tall hats, modeled on those of European soldiers, came in the 19th century. Accompanying throaty chants and a constant percussion beat,* pa'anga *seed anklets rattle over nimble feet. Elaborate flourishes of dancers' wooden paddles captivated Cook on a visit to Tonga; he thought the performance worthy of "universal applause on a European Theatre"; to give "a better opinion of English amusements," he ordered a display of fireworks that same evening.*

On January 3, 1773, after repeated sweeps in the recorded region of Bouvet's sighting, Cook decided "that what M. Bouvet took for Land ... was nothing but Mountains of Ice surrounded by field Ice." Both Cook and Bouvet were wrong. Bouvet had discovered land: a speck 5 miles wide, isolated from the nearest continent by more than 1,000 miles of open ocean. He had given its position so inaccurately that Cook was some 400 miles southeast of Bouvet Island.

Having heard of another doubtful tract, Cook sailed for the Indian Ocean to search for it. This was La France Australe, reported in 1772 by the French navigator Yves-Joseph de Kerguelen-Trémarec. En route, the crews learned something that startled them: Ice from the sea will produce fresh water. Cook sent three boats to collect ice from around a berg, and the meltwater was "perfectly well tasted," to Wales "of much more real value than Gold!"

Plunging directly south, Cook's ships crossed the Antarctic Circle—"first and only" to do so, he wrote on January 17. Pack ice thwarted them, and they veered to the northeast. Again a search of a reported location proved futile. After looking for a week, Clerke declared, "if my friend Monsieur found any Land, he's been confoundedly out in the Latitude & Longitude of it, for we've search'd ... pretty narrowly and the devil an Inch of Land is there...." Clerke was correct: The search ran 400 miles west of the desolate islands today named for Kerguelen.

"Confoundedly out" is a fair summary of longitude as navigators found it before Cook's time. Finding it was one of the grand problems—and dangers—of navigation.

Usually a competent seaman could reckon his latitude, north or south of the Equator, by measuring the height of the sun or a given star above the horizon. By the 1770s he had instruments for the purpose, and mathematical tables as well; he needed only enough fair weather to make his sighting. Longitude, distance east or west of a line of reference, was much trickier. A good navigator could find it during eclipses, which are rare; by lunars as Cook had done, which were tedious; or by dead reckoning, which was almost useless in strange waters where unknown currents affected a ship's real progress. Or he could work it out easily if he had an instrument that would keep — accurately — the time of a place of reference, such as Greenwich in England, for comparison to observed local time anywhere else. For generations such an instrument could be defined in theory, but not made. For accuracy it had to be delicate. To work at sea it had to be sturdy. Who could build such a timepiece?

A Yorkshireman managed it, after years of work: a stubborn perfectionist named John Harrison. On his second Pacific voyage Cook had a faithful copy of Harrison's fourth model, a copy made by Larcum Kendall of London. At first Cook was skeptical of this chronometer and preferred to trust his own computations. As the voyage progressed he developed respect for it. Deep in the bitter Antarctic, Cook wrote: "Indeed our error can never be great so long as we have so good a guide as Mr Kendalls watch." Cold would make its metals contract, as tropical heat would expand them; but Harrison had designed an ingenious way of compensating for this—"through all the vicissitudes of climates," as Cook proved. Still in going order today, this chronometer is preserved at Britain's National Maritime Museum. Though it resembles a plain overgrown pocket watch, it made Cook the first explorer in history who could easily find his longitude, in any weather, to an accuracy of about three miles.

Skulls peer from an opened tomb in one of Easter Island's ahu, *religious shrines noted for massive statues called* moai. *Artist Hodges found standing moai, rare at the time, sporting topknots of ruddy volcanic scoria.*

In company, the *Resolution* and *Adventure* tacked southeastward for several days; but in a thick fog on February 8 they lost contact. Each captain fired signal cannon and searched for two or three days, but to no avail. Then each sensibly decided to make the best of his way to the rendezvous: Queen Charlotte Sound in New Zealand. Furneaux angled to the northeast; Cook, always the explorer, headed to the southeast—back toward the cold.

Night after night, the watch was dazzled by the billowing glowing lights of the aurora australis. By day, the grotesque shapes of the icebergs intrigued the men, despite dangers they were beginning to appreciate. A berg some 400 feet high abruptly toppled over, close to the *Resolution.* Another shattered as if exploded. A third drew the ship toward it with a mysterious current from which she barely escaped.

After months of such sudden perils, and sustained hardships, Cook turned north and east. The last iceberg disappeared below the southern horizon, and on March 25 the familiar coast of New Zealand rose into view. The *Resolution* came upon the southwest coast of the South Island, and Cook made for "Duskey Bay," the appealing steep-walled sound he had seen at twilight three years before.

At snug Pickersgill Harbour, the weather was chilly and rainy, but the Maori were friendly, fish and wildfowl abundant and tasty, with wood plentiful and drinking water sweet. Duskey Bay, wrote Clerke, "for a Set of Hungry fellows after a long passage at Sea is as good as any place I've ever yet met with."

And it's one of the world's most ruggedly beautiful spots to explore; I was as captivated by it as Cook and his men. A long irregular arm of gray water stretches between charcoal mountains blanketed with dark green trees, and often cloaked with snow. Pewter clouds scud along the heights, scattering showers and bursts of wind. Cliffs rise from the sound, and threadlike cataracts pour into it.

Along a bouncing stream, Cook established a brewery which produced a beer concocted of spruce bark, tea leaves, molasses, and a condensed essence of malt, called "wort." Anders Sparrman, a Swedish botanist who had joined the voyage at Cape Town, happily noted: "After a small amount of rum or arrack has been added, with some brown sugar, and stirred into this really pleasant, refreshing, and healthy drink, it bubbled and tasted rather like champagne...."

Cook busied himself with surveying; his rendering of Pickersgill Harbour still appears on Admiralty chart 2589. The Forsters crammed specimens into a tiny cabin: plants, seeds, shells, birds, and fishes, which soon began to smell. Wales set up an observatory and patiently calculated latitude and longitude. A Maori family lived nearby, and Cook greeted the man politely with the ritual touching of noses. A midshipman named John Elliott wrote a sketch of Cook as diplomat: "certainly no man could be better calculated to gain the confidence of Savages. . . . He was Brave, uncommonly Cool, Humane, and Patient. He would land alone unarm'd—or lay aside his Arms, and sit down, when they threaten'd with theirs, throwing them Beads, Knives, and other little presents then by degrees advancing nearer, till by Patience, and forbearance, he gain'd their friendship. . . ."

This area explored, Cook again coasted the western shoreline. In Cook

Strait, whirlwinds stirred a series of awesome waterspouts, one of which passed within 150 feet of the ship's stern. The sky turned an ominous gray, the winds blew wildly in all directions, and the sea foamed under the spout. The "whole Atmosphere seemed in a state of Strange purterbation," Clerke wrote in wonder. With night the seas quieted, and next morning the *Resolution* reached Queen Charlotte Sound. As she edged into Ship Cove, her men saw with relief the *Adventure* riding easily at anchor.

"I my self saw a human Skeleton lying in the foundation...."

Furneaux, knowing his commander, dispatched a boat bearing fresh greens for "Salleting" and fresh fish. Then he made his report. The *Adventure* had sailed east to Tasmania, stopping for supplies before turning north. He had taken the strait between Australia and Tasmania for "a very deep bay," and Cook accepted his opinion. He had reached Ship Cove on April 7, just 12 days after the *Resolution* had arrived at Dusky Sound.

Furneaux and his men had settled in for a pleasant winter. Cook dramatically upset these plans; he meant to use the seasonal winds. As he had no desire to "Idle away the whole Winter in Port" he meant to spend the time exploring. He would sail east to the middle of the South Pacific; if no continent materialized, his ships would make for Tahiti, resupply, and then return to New Zealand, ready for a summer exploration far to the south.

"the most
wonderfull
matter
my Travels
have
yet brought me
acquainted
with."

"*Stupendous*" and "*Colossean*," Easter Island's moai topple across stony slopes and coasts. Cook's gentlemen speculated that these statues honored ancestors rather than deities —a view accepted today. At left, statues at the quarry remain embedded where they stood when work stopped abruptly in the 17th century. During the internal upheaval that followed, warring factions knocked the moai from their platforms; one fallen giant lies face up at Ahu Tongariki, the island's largest shrine (below). Carved coral and lava "eye" (far left), discovered in 1978, may once have completed a moai's face. Even more than their statues, the islanders' Polynesian origin amazed Cook and added to his respect for this far-flung people.

Again sailing in tandem, the two ships tacked out of Cook Strait on June 7. They cruised hundreds of miles south of Cook's 1769 track and then, circling north, hundreds of miles to the east. No land appeared. As they inched north into the tropics the weather became warm and pleasant, and the trade winds bore them along. On August 11 the first of the Tuamotus appeared, "these little paltry Islands" in Clerke's phrase. Cook was anxious to reach the bounty of Tahiti, because the health of the *Adventure*'s crew had deteriorated. A competent sailor, Furneaux was not the dietary taskmaster Cook was, and several of his men had fallen sick with scurvy; one man—the cook, of all people—had died.

Hearts stirred as Tahiti drew near. Men who were returning savored the memories of abundant food, enchanting countryside, and exuberant flower-scented women. Others had dreamed of such delights. All expected a sojourn in paradise.

Ticking on Greenwich time, Cook's first chronometer enabled him to fix longitude with unheard-of accuracy on his second and third voyages. It has devices to compensate for fluctuating temperatures and violent ship's motion. In silhouette stands a sextant, another new navigational aid in Cook's day.

All faced drowning before they reached shore. By some mistake, the night watch let both ships approach perilously close to a reef off the Tautira district. The wind died, and the ships drifted on the surge. In about 14 feet of water, the *Resolution* hit coral; she "continued to strike so hard that it was [with] difficulty some times that we kept on our legs," reported Wales. The *Adventure* bore down on her consort and avoided a collision by only a few feet when her anchor found bottom and held. Finally a propitious breeze gently pushed both ships from danger.

It was the priggish Sparrman who left the most telling account of the strains in such a crisis: "I should have preferred . . . to hear fewer 'Goddamns' from the officers and particularly the Captain, who, while the danger lasted, stamped about the deck and grew hoarse with shouting." Afterward, in the wardroom, Cook was "suffering so greatly from his stomach that he was in a great sweat and could scarcely stand. It was, indeed, hardly remarkable that . . . he should be completely exhausted." Sparrman prescribed a dollop of brandy—"an old Swedish remedy"—and Cook soon recovered.

All the scurvy victims soon regained their health. A shortage of pigs, and a plethora of petty thievery, inspired Cook to leave Tautira for Matavai Bay, where he was better known. Breadfruit was out of season there, and pork scarce, so he obtained little. The great ari'i Tuteha had been killed in a recent war; his successor was a timorous and inoffensive youth named Tu—in full, "Great Tu wondrous next to God"—with whom Cook established a genial relationship.

On September 1, the ships departed Matavai Bay. Cook let a youth named Porio go along, hoping he would be as helpful as Tupaia. Calm at first, Porio broke into sobs when Tahiti slipped out of sight. "Poor Porio's are not the only tears I've seen rous'd upon leaving this good Isle . . .," wrote the sentimental Clerke. "I've been in a condition myself at the time not to see a great way."

Again Cook called at Huahine and Raiatea, where he was received royally by the chieftains, entertained at lavish feasts and lively *heivas*, dramas of music and dancing. Ori, an ari'i of Huahine, greeted him like a son; he "fell upon my neck and embraced me . . . the tears which trinckled plentifully down his Cheeks sufficiently spoke the feelings of his heart." James Cook, often inexpressive among his countrymen, could quickly earn the admiration and love of islanders in Polynesia.

Now he steered west from Raiatea, hoping to confirm islands that Tasman had discovered in 1643—Eua and Tongatapu in the Tonga group. Six days out he passed two small islets enfolded by a jagged reef; he decided to name the cluster Hervey Island in honor of a Lord of the Admiralty. Isolated, jewel-like in its beauty, it now forms part of the scattered isles named for James Cook, several of which he would visit eventually.

In fair weather on a breezy October 2, the modest green highland of Eua rose on the horizon. To the northwest lay the much larger, much flatter isle of Tongatapu. These comprise the southernmost group of the Kingdom of Tonga, a diverse sprinkling of 150 islands—some atolls, some volcanoes—that Cook would come to cherish almost as much as Tahiti.

Smiling and cordial, Tongans swarmed onto the ships when they anchored at Eua. For nails, the men eagerly offered exquisite tapa cloth—and the women other favors. The astronomer Bayly wrote: "Virtue is held in little esteem here, the women gladly jumping out of their canoes & swiming to the ships sides and getting on board for the sake of a Nail or a bit of old cloth...." Second lieutenant James Burney of the *Adventure*, a young man of an artistic and well-connected family, came closer to the islanders' view: "I question if they have any Idea of Chastity being a virtue" for unmarried girls. Wives, he noted, "confine themselves to the Husband."

Cook, Furneaux, and other gentlemen were handsomely entertained by a local chief. Cook "ordered the Bag-pipes" played for the islanders; in return, the chief ordered a song rendered by three young women, to whom Cook gave a necklace apiece. Fruit was served along with kava, a heady drink made from pepper root. The Tongans chewed bits of the root, deposited these in a wooden bowl, mixed them with water, and served the liquid in cups of green leaf. "I was the only one who tasted of it," said Cook; "the manner of brewing had quenished the thirst of every one else...."

For trade in foodstuffs, Cook turned to Tongatapu. He edged along its south coast, to me one of the most impressive shorelines in the South Pacific. Unchecked by reefs, the sea incessantly pounds the coral, carving a labyrinth of bizarre formations. As each wave crashes in, the water finds small passageways in the porous rock and surges to the surface in explosions of foam that flower and fade like sparkling fireworks. Scores of these blowholes along a five-mile stretch of coast continually erupt in an eternal pageant of the power of the sea.

Although Cook spent only four days at Tongatapu on this visit, he and his men became enamored of the island and its friendly people. Green, lush, and fertile, Tongatapu still offers the lovely vistas that Cook found. Coconut palms rise high above a verdant countryside; smiling people, quick to acknowledge religious faith, offer assistance at every turn. On "delightfull Walks" through the country, Cook was captivated by its beauty: "I thought I was transported into one of the most fertile plains in Europe ... Nature, assisted by a little art, no were appears in a more florishing state...."

Taking a similar delightful walk near a small village, I came upon a group of women crouched in the shade of a large tree. With paints in subtle hues of brown and russet, they were tracing intricate designs on a piece of tapa cloth roughly 70 feet long and 12 feet wide. To make tapa, bark from the paper mulberry tree is soaked in water, arduously pounded into thin strips, then joined into sheets. Tongans still produce large rolls of tapa for all important events—births, marriages, and deaths. I noticed one girl in her teens was working with special care. Was this piece of tapa for a special occasion? She nodded. A wedding? She nodded again. Who would be married? She began to blush. "Me," she exclaimed with a grin. I kissed her cheek in congratulation, and the older women howled with laughter.

Lack of a common language barred Cook from banter like this; Porio could not understand the Tongans. Nevertheless, Cook shrewdly suspected that they spoke a "Provincial dialect" of the language he had heard from the affable Tahitians and the truculent Maori, to whose lands he now returned.

At the east end of Cook Strait, a gale struck the *Resolution* and the *Adventure* on October 23. It rose to a furious storm with "a mountainous Sea." It drove the *Resolution* far to the south, and she lost sight of her consort. On November 3 Cook finally was able to slip into Ship Cove—but it was empty. For three weeks he waited, but the *Adventure* never appeared. He decided that Furneaux had elected to make for Cape Horn and home. Before setting his own course for the Antarctic, he left a message in a bottle buried by a tree carved "LOOK UNDERNEATH." On November 27 he sailed south.

Only three days later the *Adventure*, badly mauled by the storm, limped into Ship Cove. Furneaux found Cook's note and planned to catch up with him. With repairs almost finished, he sent a boat with ten men to gather wild greens at a cove across Queen Charlotte Sound. They did not return. Next morning Lieutenant Burney was sent to investigate. All the "Indians" he encountered seemed friendly enough. He found parts of the boat, some shoes. Then: "Such a shocking scene of Carnage & Barbarity as can never be mentioned or thought of, but with horror." As well as baskets "of roasted flesh" he found two hands—young Jack Rowe's, identified by "a hurt he had received," and forecastleman Thomas Hill's, tattooed "T.H. which he had got done at Otaheite." Sickened, Burney returned to the *Adventure*; like his lieutenant, Furneaux had no stomach for revenge. With all haste he made sail, crossed the Pacific, searched desultorily for Cape Circumcision, and reached England in July 1774.

The *Resolution*, meanwhile, was hauling steadily south, to cross the Antarctic Circle. She found herself in snow and fog and "confoundedly entangle'd with loose and field Ice," as Clerke reported. Poor ship handling by the officer of the watch brought her into the path of an iceberg twice the height of her masts; she evaded it by inches—"the most *Miraculous* escape from being every soul lost, that ever men had," wrote Midshipman Elliott. Appalling weather forced Cook northward. "Our ropes were like wires," he wrote on Christmas Eve, "Sails like board or plates of Metal . . . the cold so intense as hardly to be endured, the whole Sea in a manner covered with ice, a hard gale and a thick fog. . . ." The Christmas festivities had grim overtones that year. As Sparrman recalled, the "Sailors and marines . . . joked about the voyage, and vowed that, if they were wrecked on any of the 168 masses of ice surrounding us, they would certainly die happy and content, with some rescued keg of brandy in their arms."

As 1774 began, Cook was steering the first curve of a U-shaped swing that took him 1,500 miles north—into the very heart of Dalrymple's continent—and then back to the Antarctic Circle. His men were eating repulsive salt meat by now, and spoiled bread. He plunged on south, farther south than ever before, and indeed farther south than anyone else would reach for nearly half a century. On January 30, snow-white clouds on the southern horizon reflected the brilliance of ice ahead: a mile of pack ice, with solid ice beyond rising to ice mountains.

It was now he penned that rare self-revelation: "I will not say it was impossible anywhere to get in among this Ice, but I will assert that the bare attempting of it would be a very dangerous enterprise and what I believe no man in my situation would have thought of. I whose ambition leads me not only farther than any other man has been before me, but as far as I think it possible for man to go, was not sorry at meeting with this interruption, as it in some measure relieved us from the dangers and hardships, inseparable with the Navigation of the Southern Polar regions."

For once, pride resounds in his words. The understatements of his summary follow day-by-day details of risk incurred on a ship alone in a region that—unlike Cape Tribulation—precluded survival for any castaway. With this exploit James Cook had truly inscribed his name among the luminaries of discovery, a new star in the constellations seamen steer by.

He was not yet ready to go home. If he had refuted all theories of a continent in the Pacific, it still held "room for very large Islands" undiscovered, while many of those recorded were "very imperfectly explored and there situations as imperfectly known." In the next nine months he discovered or rediscovered dozens of islands, dredging many from the shadowy fringes of history. By December 1774 virtually every important island group in the South Pacific had been located, identified, and charted.

Illness—nearly fatal—struck him almost before he could begin. He tosses off his sickness in few words: "I was now taken ill of the Billious colick and so Violent as to confine me to my bed." George Forster wrote "that his life was entirely despaired of." The captain suffered an intestinal blockage complicated with vomiting and wracking hiccups that lasted almost 24 hours. Surgeon James Patten plied him with purges and emetics, opiates and glysters—all to no avail. Finally hot baths and stomach plasters brought him ease. Some doctors today diagnose his malady as an acute infection of the gall bladder with secondary paralysis of the intestine; others think he was infested with parasites swallowed in some native food.

He was far from well as the *Resolution* approached his next objective, Easter Island. A forlorn 62-square-mile speck in the eastern Pacific, it had been discovered by the Dutch explorer Jacob Roggeveen in 1722. Around it at each compass point stretch a thousand miles of empty ocean, an ocean that incessantly shatters the rocky volcanic coasts with powerful breakers. The chill winds that sweep these immensities tested the ships of the 18th century—and effectively marooned the ancient seafarers who landed here after some untraceable epic voyage. Now the winds seem to carry a brooding melancholy that pervades the island and its present 2,100 inhabitants.

"Here the Polynesians had developed an extraordinarily sophisticated culture—where you'd least expect it," said Chilean archaeologist Gonzalo Figueroa. An elegant, intellectual man with snow-white hair, he chatted with me in the sitting room of a cozy guest house in Hanga Roa, the only modern settlement. "The first settlers faced a challenging environment: a colder climate, limited wood and little water. And above all, cultural isolation. They turned inward, to religious expression."

That expression took an imposing form in the celebrated statues called *moai*. These are enormous. Most stand about 20 feet high; they weigh as much as 80 tons. Somehow they were moved across miles of rolling treeless countryside to the altars of scattered communities. The labor of it strains the imagination. "These islanders," said Gonzalo, "had an obsessive dedication to their religion—to their rituals."

Scores of *ahu*, or open-air temples, dot the coastline; the mute, unnerving stare of the moai reveals nothing of their purpose. Hundreds of them lie partially finished in the volcanic quarry or abandoned between the quarry and the ahu, as if catastrophe had ended the work.

Excavating an ahu by a beach called Anakena, archaeologist Sergio Rapu took the time to discuss these riddles with me. Born on Easter Island and trained at the universities of Wyoming and Hawaii, he has a keen awareness of the many factors involved. "There is no sure answer to anything about Easter Island," he said with a shake of his head. "But we have some good data, and we can offer theories. Our traditions tell us that originally two canoes, with a total of some three hundred people, landed here about A.D. 400. Our first confirmed radiocarbon date is A.D. 690. By the 17th century the population had expanded greatly—maybe to 10,000. The island could not support them. Also, the social and political systems, workable for a smaller society, would come under strain. Perhaps at the same time the island suffered one of the severe droughts that periodically strike us."

From his discussion I could picture crisis evolving into chaos. One group would raid another for food and water; those raided would seek revenge, stealing food as well as toppling and desecrating the moai. Violence would spiral into killing. Food supplies would dwindle until people turned to cannibalism. "What was once an elevated and tightly controlled society had broken down, almost to anarchy," said Sergio. "The first Europeans arrived after the worst phases of destruction, but disorder still reigned."

Bewildered by many aspects of the society, Cook and his men were awestruck by the huge statues and the labor they implied; to Clerke this achievement was "the most wonderfull matter my Travels have ever yet brought me acquainted with." The islanders were "exceedingly civil" in manner, but played low-comedy variations on thievery. They had a passion for hats, which they snatched off their visitors' heads. They had tasty produce—bananas, sweet potatoes, yams—but not much of it; they traded off their neighbors' crops, they demanded payment in advance and instantly fled with it, and under top layers of good potatoes they hid makeweight stones. After only three days, with his crew showing early symptoms of scurvy, Cook decided to make for an island group 2,300 miles northwestward, the Marquesas.

While his men were wondering if a "race of giants" had made the statues— or grumbling because here the men had kept their women hidden—Cook was comparing the Easter Islanders with people of the "more Western isles." In physique and language they bore "such affinity" as to prove a common origin. He marveled at their dispersion "over all the isles in this Vast Ocean ... almost a fourth part of the circumference of the Globe." With striking insight, he reasoned that they had "by length of time become as it were different Nations" with different customs. His wonder—and respect—would grow as he continued to pursue the Polynesians across the Pacific.

The Marquesas, today a remote outpost of French Polynesia, were discovered by the Spanish nobleman Alvaro de Mendaña and his Portuguese pilot Pedro Fernandez de Quiros in 1595. Wildly breathtaking in their beauty, these islands jut abruptly from the sea and soar into the clouds. Molded of dark volcanic rock, eroded into fantastic escarpments, draped with dark green vegetation, they are bold and imposing. Their beauty and the statuesque loveliness of their inhabitants have enticed artists for many years: the American novelist Herman Melville, the French Impressionist painter Paul Gauguin, and more recently Jacques Brel, singer and composer of French songs. Both Gauguin and Brel are buried on Hiva Oa, largest of the group.

With Mendaña's account in hand, Cook cruised the eastern coast of Hiva Oa, ran the channel between it and Tahuata; and anchored in the bay named Madre de Dios by the Spaniard 179 years earlier. Two hundred and five years after Cook, Gordon and I followed the same track, but in a small open boat awash with spray. The waves, unchecked by land clear from South America, loomed above us as we slid into troughs between them, then carried us high toward their towering crests. A pod of porpoises frolicked alongside, their silvery bodies streaking past, bursting into the sunny air, then belly-flopping back to the water. (Continued on page 129)

"Without exception the finest race of people in this Sea," Cook pronounced the handsome Marquesans. His ancestors' legacy lives on in a young man of Hiva Oa.

Volcanic cliffs of Fatu Hiva plunge into Hanavave Bay. Beset by surf, squalls, and
a trading fiasco on nearby Tahuata, Cook bypassed Fatu Hiva as he made for Tahiti.
Today the isolated Marquesas rank among the least developed Polynesian isles.

*"the Hills Of the Marquesas in most places
break off perpendicular into the Sea...."*

"the Volcano...
vomits forth vast
quantities of black
Smoke, and
in the Night
flames of Fire
are sometimes
seen."

In predawn light, a village guide climbs to the rim of Yasur Volcano on Tanna, in the Melanesian group Cook named the New Hebrides. "A great fire" guided Cook to the island, but Tannese offering a tour of the volcano led his men, instead, "down to the harbour before they preceived the cheat."

After two exhilarating hours, we reached the deep horseshoe of the bay—called Resolution by Cook, today officially Vaitahu — on the precipitous island of Tahuata. At last Cook expected to nourish his ailing crew, deprived of fresh provisions for virtually five months. He began exchanging the usual nails, spikes, and hatchets for the usual hogs, fruits, and vegetables. Trade was brisk until one young gentleman, greedy for a particularly fine pig, offered red feathers from Tonga. Suddenly the market disintegrated; the Marquesans desired only the red feathers. Cook was incensed that his rules had been broken, a "fine prospect" of supply frustrated.

Wearily, Cook weighed anchor and set his course again for Tahiti. Everyone who kept a journal was recording his admiration of the handsome Marquesans — who had kept their women at a prudent distance. Wales thought them "almost without exception all fine tall stout-limbed, and well made People." To Clerke, the heavily tattooed men were "all in fine Order & exquisitely proportion'd," while the women were "remarkably fair" and "very beautifull" with long hair worn "down their Backs in a most becoming and gracefull manner."

More dispassionately, Cook noted that a Tahitian could "converse with them tolerable well" — another sign of affinity. Indeed, deep in the thick forests of Puamau, a valley on Hiva Oa, Gordon and I came upon squat stone statues in an ancient and overgrown arena. Archaeologists now theorize that the Easter Islanders emigrated from the Marquesas and that these weathered monuments are the counterparts of the famous moai.

Bypassing the romantic island of Fatu Hiva, southernmost of the Marquesas, Cook began a pleasant journey southwestward. For the third time he cautiously tacked through the Tuamotus, to anchor in well-loved Matavai Bay. As always, the English were greeted with flourishes of friendship; old loves were rekindled and new ones ignited. In sharp contrast to his previous visit, Cook found food in abundance, new houses and canoes "built and building," and other signs of peace restored.

He had planned a sojourn of two or three days, to let Wales check any error in "Mr Kendalls watch" against well-known latitude and longitude; he lingered for three full weeks. The familiar joys, confusions, and problems ensued, and he was forced to inflict punishment on crewmen and natives alike. He had held his hand as long as he could. Perhaps the spectacle of a Tahitian naval review — 330 double canoes, "very well equip'd, Man'd and Arm'd" — lingered in his mind as he analyzed his relations with the islanders in an almost prophetic statement. "Three things made them our fast friends, Their own good Natured and benevolent disposition, gentle treatment on our part, and the dread of our fire Arms; by our ceaseing to observe the Second the first would have wore of [f] of Course, and the too frequent use of the latter would have excited a spirit of revenge and perhaps have taught them that fire Arms were not such terrible things as they had imagined, they are very sencible of the superiority they have over us in numbers and no one knows what an enraged multitude might do."

Sailing southwestward, Cook reached a high jagged isle that he decided to investigate. At the first landing, Sparrman was struck with a stone and

Low tide uncovers a steaming volcanic spring at Port Resolution. Cook's men discovered Tanna's hot springs after one sailor burned his fingers while gathering stones for ballast. In Tannese lore, Cook stole the island's sacred stones of wisdom to give Europeans the power and knowledge they now enjoy.

PRECEDING PAGES: *Traditional fishing methods survive on Tanna, where the sparkle of coconut-frond torches lures reef fish into range of arrows and spears.*

two men fired without orders; the "Indians" fled, and Cook decided to try again down the coast. He went ashore with two boatloads of men, to leave gifts in some deserted canoes; his party was promptly attacked, one native charging "with the ferocity of a wild Boar." Spears whistled past Cook's head. Musket fire scattered the attackers, and Cook quickly returned to the *Resolution,* thinking that "no good was to be got of these people or at the isle. . . ." He named it Savage Island; today it is called Niue.

Courtesy—and light-fingeredness—marked his reception at Nomuka, one of the Tonga Islands discovered by Tasman. Yams, coconuts, and grapefruit were liberally offered for trade. And generosity went still further. On the first morning, as Cook stood on the beach, he was approached by an elderly woman who offered him a ravishing young girl in return for a nail or a shirt. Cook declined, pleading poverty. He "was made to understand I might retire with her on credit, this not suteing me niether the old Lady began first to argue with me and when that fail'd she abused me. . . ." Her actions explained the words Cook could not understand: "Sneering in my face and saying, what sort of man are you thus to refuse the embraces of so fine a young Woman, for the girl certainly did not [lack] beauty which I could however withstand, but the abuse of the old Woman I could not and therefore hastned into the Boat. . . ." For once, Commander James Cook, R.N., indomitable in so many adverse situations, was reduced to flustered retreat. Midshipman Elliott penned a comment on Cook's conduct in such matters: "It has always been suppos'd that Cook himself, never had any connection with any of our fair friends; I have often seen them jeer and laugh at him, calling him Old, and good for nothing."

Despite this awkward encounter and several serious incidents of theft, Cook at his departure had settled on a name for the entire island chain: "this groupe I have named the Friendly Archipelago as a lasting friendship seems to subsist among the Inhabitants and their Courtesy to Strangers intitles them to that Name." Even today, the Tongans are proud of that heritage; every person I passed on a road or encountered in a town smiled and waved a greeting—especially the children.

Now generally pleasant weather and gentle winds transported the *Resolution* toward another group improperly located on the map and only partially explored. In 1605 the Portuguese navigator Quiros had mounted his own expedition across the Pacific, hoping in unbounded zeal to found a colony and convert the heathen to Christianity. He discovered what he thought was a continent, and planned to build the utopian city of New Jerusalem. Illness and untrustworthy officers conspired against him, and he returned to Mexico a broken man. Bougainville, after visiting Tahiti in 1768, had sailed the same waters, identifying several islands but never confirming the truth about the "continent."

This was left to Cook. He sighted land just about where Bougainville did, on July 17, 1774; for the next six weeks he established the character of the volcanic group that he named the New Hebrides. He executed an admirable running survey of the chain—a scattering of a dozen and a half chunks of land strewn across some 450 miles of ocean. The terrain he found was dramatic and steep. Several islands were aglow with active volcanoes. And the people were strikingly different from the Polynesians. There was none of the noble handsomeness of the Marquesans, the puckish friendliness of the Tahitians, the fierce intelligence of the Maori. On first impressions Cook called this Melanesian people a rather "Apish Nation."

His first encounter ashore—like each succeeding one—left him puzzled and frustrated. At sunset on July 21, the *Resolution* came to anchor in a bay of the large island of Malekula. Next morning, bearing a green branch to signify peace, Cook waded ashore, hoping to establish trade and get firewood. Confronting him were several hundred men armed with clubs, spears, and bows and arrows. Cook distributed medals and tried to speak to them in all the phrases he knew. In return, he received a token present: one small pig, half a dozen coconuts, a little water. The Malekulans showed no interest in metal objects; all they wanted, Cook concluded, was for the visitors to be gone. Recognizing futility, he acceded to their wishes.

Noting the distinctive traits of these people, Cook sketched their appearance: "almost black or rather a dark Chocolate Colour, Slenderly made, not tall, have Monkey faces and Woolly hair. . . ." For clothing the men wore only penis sheaths, the women a kind of skirt. For cosmetics men used a black paint, women a yellowish red like turmeric dye. Young George Forster thought that their features, "remarkably irregular and ugly, yet are full of great sprightliness, and express a quick comprehension." All the Europeans found their language utterly strange and hard to pronounce, but Cook "observed that they could pronounce most of our words with great ease."

When he tried to land on Erromango, he had even less success. Again he walked up the beach with a green branch to an armed but apparently courteous throng. He distributed a few trinkets. A "chief" presented a sparse gift of food and water, then gestured for the longboat to be pulled onto the beach. Suspicious, Cook stepped back into the craft and ordered his men to row offshore. The Erromangoans grabbed at the boat and oars to drag it back. Cook wrote sadly that "our own safety became now the only consideration. . . ." It became "absolutely necessary for me to give orders to fire as they now began to Shoot their Arrows and throw darts and Stones at us, the first discharge threw them into confusion. . . ." Four of them fell.

One seaman had been nicked in the cheek by a spear, one bruised by a spent arrow. Cook sailed on, naming a nearby bluff Traitor's Head.

What Cook could never know was that the people of the New Hebrides saw him as an uncanny apparition. "They lived," explained Professor Beaglehole, ". . . on the narrow and terrible border of the unseen. They were very close to their dead; they feared the spirits of the dead, the menacing and maleficent ancestors. . . . It was the part of wisdom to avoid or propitiate ghosts; but if the worst came to the worst they could be attacked with human weapons and driven away. . . . Spirits, ghosts, were not the colour of human beings, they were white. Cook and his men were white." An Erromangoan tradition says that the islanders believed they had repelled an onslaught from the supernatural, for the ship vanished into thin air.

In growing need of wood and water, Cook ran with the wind to the south, toward a "great fire." This turned out to be "a Volcano which threw up vast quantaties of fire and smoak and made a rumbling noise which was heard at a good distance." Nearly in the shadow of Yasur Volcano, Cook stopped in a narrow bay on the island of Tanna. "Vast numbers of the Natives" gathered on the beach to watch; some armed men ventured close in canoes. One elder paddled out to the *Resolution* with "2 or 3 Cocoa nutts or a yam," duly paid for with trade goods. Encouraged, the old man made several more trips.

Landing with three boats of armed marines and seamen, Cook faced two groups of men also armed—a thousand and more. On the sand lay a meager offering of food. Cook gestured for the warriors to move back; they held their ground even after a volley of musket fire rang over their heads. Alarmed for a moment, "they recovered themselves and began to display their weapons, one fellow shewed us his back side in such a manner that it was not necessary to have an interpreter. . . ." Cook signaled for the *Resolution* to fire her cannon over the throng, which indeed dispersed except for one man— the elder, named Paowang, who had brought produce to the ship.

Cook had his men stake out lines along the way to a freshwater pond, and carefully stay between them as they filled the ship's casks. With gestures he won Paowang's permission to cut firewood. Perhaps because of Paowang's example, the islanders came drifting back. Far from stealing anything, they seemed afraid to touch the Europeans' possessions. They never let the strangers go far inland, but did let them gather stones for ballast near the beach. During the two August weeks Cook spent at Tanna, an uneasy accommodation prevailed, with Paowang a constant friend.

*U*ndaunted by "ghosts" in Cook's day and by Europeans since, the Tannese today keep a fierce pride in their own ways. One blisteringly hot afternoon in September, Gordon and I hiked deep into the humid jungle to witness a boys' circumcision ceremony. A man of any one village must marry a woman from elsewhere, and his village must supply a woman to take her place sooner or later; so this important ceremony brings together people from a distance, linked by ties of long standing. We found a hundred people hunkered in a circle in a dusty clearing. The sun baked the bare earth, and motes of dust drifted in its rays. Some of the people wore traditional dress; faces were painted in swirls of red and yellow.

Attention was focused on the center of the circle. Here the father of each boy initiate would receive gifts from men of the mother's family, and offer gifts in return. Large bundles of *laplap*, yam or taro paste wrapped in green leaves, changed hands. Misshapen roots of kava were piled high. Fat hogs, their legs bound tightly, screeched mightily until their skulls were crushed by blows of a club. Emaciated dogs snuffled at the blood. When the exchanges were complete, the men withdrew to the edge of the circle. Soon a file of other men, their faces whitened, escorted in the boys who had been circumcised. While their wounds healed they had been kept in isolation, with older men to attend them. Now they were being presented to their parents as men. Under their feathers and paint, the boys — apparently ranging in age from eight to twelve—looked dazed by their experiences, now to culminate in a great feast and joyous dancing long into the night.

"I can't imagine that rituals like these have changed much at all since Captain Cook's time," remarked Gordon as we returned to the trail. Here, as nowhere else, we could appreciate the shock of that first meeting of alien worlds. This rendered all the more impressive Cook's effort to put himself in the islanders' place: "its impossible for them to know our real design, we enter their Ports ... we attempt to land in a peaceable manner, if this succeeds its well, if not we land nevertheless and mentain the footing we thus got by the Superiority of our fire arms, in what other light can they than at first look upon us but as invaders of their Country. . . ." In July 1980 the New Hebrides, formerly under the joint rule of France and Great Britain, became the new nation of Vanuatu; and Cook if anyone could appreciate the difficulties it might face. Already, with unhappy insight, he had assessed the damaging impact of Europeans upon the peoples of the Pacific: "we interduce among them wants and perhaps diseases which they never before knew. . . ." The moral perspective he brought from his parents' home and the parish church in Marton-in-Cleveland did not fail him in the unexpected rituals of Oceania.

After restoring his supplies of wood and water, Cook sailed from the Tannese harbor that he named Port Resolution. For 11 days he looped through the New Hebrides, placing them on his chart. On September 1, the *Resolution* once again was cutting through untracked ocean. Cook meant to reach New Zealand as soon as possible, *(Continued on page 138)*

Her headdress denoting status, her face reflecting concern and pride, a Tannese mother awaits her son's return after the long seclusion that follows circumcision.

*T*ransformed by circumcision
and ritual ordeals, boys return
as adults to their celebrating
village. Men who attended the
boys during their seclusion join
the procession (right), and feed
an initiate, who may not touch
his own food (below, right).
Other villagers haul a pig in for
the feast. Valued beyond all else,
pigs confer power on their
owners. One bound specimen
takes its place in an assemblage
of kava plants and other items
(below, center) that will serve as
gifts for the boys and exchanges
to honor matrimonial ties with
other villages. Kava, as gift or
potion, sanctions all important
events from birth to death.

"they
like wise
gave us
to understand
that
Circumcision
was practised
amongest
them."

Fringed proud eyes watch Tanna's age-old initiation rites. In "what other light can they

"We found the
Natives here
far less
friendly...."

than at first look upon us," mused Cook, "but as invaders of their Country. . . ."

resupply, and take advantage of the southern-hemisphere summer to explore the high latitudes before returning to England. But he still had "some time left to explore any lands I might meet with." One such land appeared on the horizon within three days.

New Caledonia, some 250 miles long and 30 wide, cuts the Pacific like a long finger pointing a course. On a tropically warm morning, I stood on the sandy fringe of Cape Colnett — named for the midshipman who first sighted the island — and gazed inland. A band of trees gave way to brush-covered hills, which in turn yielded to grassy mountains rising into the clouds. Behind me, huge Pacific breakers exploded on the offshore reef that encircles the entire island: the most dangerous coast yet in this voyage.

Finding a break in the reef, Cook made his landing on September 6. The Melanesian crowd that gathered round him proved shy but friendly — a pleasant change from the hostility of the New Hebrideans. These people, said Cook, "had little else but good Nature to spare us. In this they exceeded all the nation[s] we had yet met with, and although it did not fill our bellies it left our minds at ease."

"Here the Trees & Verdure are of a fine bright Green...."

By sheer luck Cook and the Forsters escaped death two days later. They tasted the liver and roe of a large and ugly fish, which proved to be poisonous. They suffered extraordinary sensations of weakness and numbness, relieved only by emetics and induced sweating.

Recovered, Cook traced the entire eastern coast, always staying prudently outside the reef. At the southern tip he ran great risk to examine its bizarre trees, but hard gales kept him from making headway along the southern coast. Reluctantly, he left it unexplored and made southward. His uncanny sense for land brought him soon to tiny Norfolk Island, where he discovered the elegant pine trees which have made that island famous.

On October 19 the *Resolution* lay at rest in Ship Cove. In three weeks of furious activity the men were stuffed with greens and the ship was overhauled for the voyage home. She left Cook Strait on November 10 and with friendly winds reached the west coast of Tierra del Fuego in just five weeks. Cook duly explored this storm-laced coastline for eleven days, calling it the most barren yet, and spent the third Christmas of this voyage in a safe but desolate harbor. Some carefully hoarded Madeira wine warmed a dinner of roast goose, boiled goose, and goose pie. It stirred the captain to one of his rare jokes—he described the Madeira as "the only Article of our provisions that was mended by keeping; so that our friends in England did not perhaps, celebrate Christmas more cheerfully than we did."

Four days later the *Resolution* gained Cape Horn. "At half past 7," Cook wrote, "we passed this famous Cape and entered the *Southern Atlantick Ocean.*" There he would vainly search the icy waters for signs of Dalrymple's moribund continent. On July 30, 1775, three years and 18 days after departing, he would anchor again in English waters. But on December 28, still poised off Cape Horn, he gazed back at the Pacific Ocean—perhaps with a touch of wistfulness. He was leaving it forever—or so he thought.

Scenes of New Caledonia: On tiny Améré Island, tall araucarias, sometimes called Cook pines, commemorate their discoverer. Poling his bamboo raft along the placid Tiwaka River, an islander heads home in softening daylight (right).

The Third Voyage:
"The Call of My Country for More Active Service"

*L*ionized by society, esteemed by the scientific world, praised by the King, James Cook returned to London and a tumultuous reception. His fame spread through the kingdom, and he was acclaimed wherever he went. George III eagerly heard his report and approved another promotion —to post captain. The Royal Society unanimously elected him a Fellow. He dined at select clubs and at the homes of the prominent, where he met such notables as man-about-town James Boswell, who found him "a plain, sensible man . . . very obliging and communicative."

"At this point, Cook had risen to the pinnacle of his career," commented historian Michael Hoare. "I like to compare him to an American contemporary of his—Benjamin Franklin. Although they pursued disparate careers, their lives are strikingly similar. Both were born to an humble station in society, but early on showed a degree of intelligence and perception that seemed to mark them for something great. They worked hard, and learned quickly how to play the establishment in their favor. They showed more and more ability with each new position they assumed, and each reached the top of his chosen field. Of course the chips all had to fall right, too, but both Cook and Franklin made lasting and incomparable contributions to the world."

The Lords of the Admiralty rewarded Cook by appointing him to the board of the Greenwich Hospital, a sinecure that offered excellent pay and abundant free time. This surely pleased Elizabeth Cook and young James and Nathaniel, deprived of husband and father for six of the previous seven years. Tragically, Mrs. Cook had lost another infant son three months after the *Resolution* had sailed; now, however, she was pregnant again.

Cook accepted his new appointment, but on condition. He obtained assurance that he might "quit it when either the call of my Country for more active Service, or that my endeavours in any shape can be essential to the publick." To his old friend John Walker he confided: "my fate drives me from one extream to a nother a few Months ago the whole Southern hemisphere was hardly big enough for me and now I am going to be confined within the limits of Greenwich Hospital, which are far too small for an active mind like mine, I must however confess it is a fine retreat and a pretty income, but whether I can bring my self to like ease and retirement, time will shew."

Time, indeed, did show. Already rumors of another Pacific expedition were circulating. The *Resolution* was drydocked to be refitted for it. But Cook did not expect to command it. After years of battling the elements, of eating dubious foods, of enduring the strain of responsibility, he was weary. He was 47 years old, and he had been at sea almost constantly for 30 years. Instead of relaxing, however, he busied himself with his logs and journals, composing the official account of the second voyage. A journalist named John Hawkesworth had produced the chronicle of the *Endeavour* voyage, and Cook was aghast at its inaccuracies. He told James Boswell that Hawkesworth would draw "a general conclusion from a particular fact, and would take as a fact what they had only heard. . . ." Boswell—used to distinctions of genius from his friend Dr. Samuel Johnson — decided that Cook "had a ballance in his mind for truth as nice as scales for weighing a guinea." Neither an experienced nor a particularly gifted writer, Cook spent grueling hours in the fall and winter of 1775, rewriting and polishing his journals.

His life changed course at a party in early 1776: Sandwich invited him to dinner with Sir Hugh Palliser and Sir Philip Stephens. Cook's first biographer, Andrew Kippis, who probably heard the story from the noble lord himself, wrote that everyone wanted Cook for commander and no one

PRECEDING PAGES: *Dancers of Aitutaki in the Cook Islands perform at a local fete with the verve, precision, and grace that impressed the English explorers. Cook visited some isles of this far-flung group on his second voyage, others on the third.*

dared ask him to run such risks again. The men of the Admiralty discussed the plans—and grandeur—of the project until "Captain Cook was so fired . . . that he started up and declared that he himself would undertake the direction of the enterprise." His hosts' "secret wishes" were secured.

Thus, less than six months after coming home, Cook had committed himself to a third major voyage. He had also undertaken to settle a great geographical question: the existence of the notorious Northwest Passage, a sea route between the Atlantic and Pacific Oceans, giving Europe easier access to the riches of the Far East. Since John Cabot's voyage from England in 1497, many explorers had searched the North Atlantic coast in vain for this channel. A few adventurers had prowled the storm-pounded Pacific shores with equal lack of success. No one, however, had completed a search of that

coast. Cook was assigned to accomplish this. If he found a passage to "Hudsons or Baffins Bay," he should use his "utmost endeavours" to make it his route home.

In addition, he was to return to his homeland a youth from Huahine, Omai, whom Captain Furneaux had brought to England on the *Adventure*. Under the protection of Joseph Banks, Omai played the part of the noble savage with great success in the best society. He appeared at court in velvet and lace, saluting His Majesty with "How do, King Tosh!"

Eager to benefit the peoples of the islands, the King graciously sent them garden seeds and livestock, with more to be bought at the Cape of Good Hope. The duty of providing water and fodder increased his officers' worries, while his sailors faced the added chores of cleaning the stalls.

Yuletide of 1776, early in the third voyage, finds Resolution *and* Discovery *far south in the Indian Ocean at uninhabited Kerguelen Island, anchored in Christmas Harbour. Penguins, seabirds, seals, and meager greens filled out the holiday meals. For the "stirility" of this cold, rocky, treeless spot, fog-shrouded day after day, Cook proposed calling it "Island of Desolation."*

Busy with his writing, stealing time to sit for a portrait, Cook saw to the purchase of still another cat-built bark: the little 300-ton *Discovery*. Affable Charles Clerke proudly assumed command; a disciple of Cook's, he was much better matched to his captain than Furneaux had been. James Burney was his first lieutenant; John Gore, the kangaroo hunter of the first voyage, was Cook's. Young James King, genteel and discerning, was Cook's new second lieutenant. John Webber was appointed as artist; various officers, including the inspired surgeon William Anderson, assumed the roles of naturalists. Two young men, both to become immortal in the annals of the sea, leaped at the chance to sail with Cook. George Vancouver, explorer in the making, signed on as midshipman; and William Bligh, later captain on the mutinous *Bounty*, became master of the *Resolution*.

Cook had planned to sail in April, but various delays kept him in port until July. His account of the second voyage appeared later, in May 1777, to wide acclaim. He introduced his work as "for information and not for amusement," and apologized for lacking both "the advantage of Education" and "Natural abilities for writing." His disclaimer was unnecessary; his account remains a classic in the literature of discovery.

In the last week of June, Cook bade his family—including newborn Hugh—a final farewell. He took the *Resolution* to Plymouth to join the *Discovery*, and saw a war convoy on its way to help subdue the rebellion in America: "unhappy necessity." At Plymouth he learned happier news, of an award from the Royal Society. The Copley Medal, Britain's highest honor for intellectual achievement, was conferred on him for his success in checking

"and there Are some Verry Pretty Lasses Here I'll Ashure You...."

"Pretty Enuff"—so American-born Lt. John Gore called the floral garlands worn in the southern Cooks. Mareta Anaterea adjusts a wreath of freshly picked frangipani for her friend Teeiau Maeva; at right, they stroll along a white-sand beach on Tapuaetai, an islet in the fringing reef of Aitutaki. "For the islanders, this is a favorite picnic spot," says the author; "for me, it's the loveliest place I saw in all the Pacific. Cook sailed by at night, some miles off to the west; I think if he had stopped here, even he would never have left!"

scurvy, as described in a paper he had read to the Fellows in 1776. I had the pleasure of examining its six crisp pages in the dignified, high-ceilinged library of the Society. In bold handwriting, Cook had detailed his uses of "Sour Krout" and scurvy grass, of fresh fruit and spruce beer — and his firm methods of application. He must have been elated at his prize, although he wrote politely and modestly to Banks of "this unmerited Honor."

Finally, on July 12, 1776, the *Resolution* slipped her moorings and headed southwest — but without the *Discovery*. Charles Clerke had been thrown into prison, having guaranteed debts defaulted on by an insolvent brother. In three weeks he gained his release from the gaoler's clutch, but he never truly gained his freedom—in his lungs lurked the dreadful scourge of consumption.

At Cape Town the two ships joined company. On this voyage Cook had paused at Tenerife in the Canary Islands to buy feed and straw for the stock and wine for the people. Although the price was better than at Madeira, he was chagrined to find the wine much inferior. Worse, both ships were suffering from shoddy work in the Navy's drydocks. Poor caulking led to leaks; surgeon Anderson noted the menace to health—"few could sleep dry in their beds." Inferior rigging and sails would plague both ships the entire voyage, and rotten masts would put them in mortal danger. Corruption, a long-standing abuse, had flourished under Sandwich. Cook in person had held earlier preparations to the honest standards of Whitby; this time he had left supervision to others while he trimmed and spliced his journals.

He ordered repairs at Cape Town. He

(Continued on page 153)

*Y*oung men of Mangaia, southernmost of the Cook Islands, pass an afternoon on
the reef that lies about a hundred yards offshore, fishing in deep water beyond it.
One grips a bamboo rod strained by a big fish —or a hook snagged in the coral.
Polynesian fishing gear included nets, traps, spears, and bone hooks used with a
handline; casting for sport came with later visitors. Cook, scouting the coast of
Mangaia in March 1777, saw surf breaking "with great fury" against this reef.
Finding no safe anchorage among the sharp coral boulders, no safe landing place,
and no cordial welcome, he regretfully sailed on northward from "this fine island."

*L*impid shallows of Aitutaki lagoon reveal coral not as a ship-mangling hazard but as a living wonder — "one of the most enchanting prospects that nature has any where produc'd," wrote Cook's eloquent surgeon-naturalist William Anderson. Here grayish-pink algae cover a limestone ridge; branches of an Acropora species reach toward gray brain coral; and polyps of a Montipora species show a vivid magenta. Cook's men saw similar vistas at Palmerston Atoll, then uninhabited. Lacking anything like modern snorkeling masks or diving gear, they examined the reef areas closest to the surface. Inspecting these, they marveled at the brilliant colors — yellows, blues, reds, and blacks — of fishes "playing their gambols amongst the little caverns." Probably they never glimpsed squirrelfish like those opposite, which usually hide in the coral by day and venture out at night.

"the Eye
could never tire
but view
every spot with
fresh transport,
and wonder
for what purpose
Nature should
want to conceal
a work so
elegant...."

"They have a stout boom, which they use
as an outrigger...this, when it blows fresh & the boat heels,
some of the Men go out & stand upon."

*H*eeling sharply, a sailing canoe skims across Aitutaki lagoon as her owner restores her trim. Left hand on the tiller, he throws his weight to starboard to bring the outrigger down to the surface. Lt. James King described an "exceeding narrow" outrigger canoe, "neatly made," that ventured close to the Resolution off Mangaia. Small craft like this served well in sheltered waters; the long Polynesian voyages required larger, sturdier vessels with double hulls.

increased his menagerie, buying rabbits and poultry as well as rams, ewes, goats, and horses. Omai gave up his cabin to two mares and two stallions. With water and provender crammed aboard, Cook left the Cape on December 1 in tandem with Clerke. He decided to search again for the islands Kerguelen had reported, and this time he had a more accurate position for them. Weather was against him, however; Clerke noted that "the confounded foggy Atmosphere renders exploring a miserable business." Despite its hazards, Cook soon came upon the elusive isles—as barren as any he had ever found. From Christmas Eve to December 30, he scouted bleak harbors and rocky capes. Although flowing streams provided water, there was no wood and little grass for the stock. Thomas Edgar, master of the *Discovery*, wrote of the incessant "Melancholy Croaking of Innumerable

Penguins," and Gore was more than happy to leave "this Cold Blustering Wet Country."

On January 24 the southern coast of Tasmania came into view, and soon the two ships were comfortably anchored in Furneaux's Adventure Bay. A broad, gentle bight, it keeps its pleasing aspects even now. I sipped cool water from the stream where Cook had refilled his casks; it chuckles through shady glens before meandering across white sand and pouring into the bay. Small parties of natives, wearing neither clothes nor ornaments, greeted the strangers. "Upon the whole," wrote Anderson, "these indians . . . seem mild and chearfull." The impact of Europeans on the peaceable Tasmanians proved calamitous; they were extinct by 1876.

Singing, strumming a ukelele, a musician of Aitutaki follows a beat set by drums as men present the vigorous dances that evoke canoe voyages or challenges to battle.

Perhaps because he was some weeks late under his planned timetable, Cook did not test Furneaux's notion that Tasmania formed part of the mainland of "New Holland." More than twenty years would pass before the truth was known. He raised New Zealand by February 10, and made for his anchorage in Ship Cove. With the marines on guard against attempted massacres, his men worked armed —repairing the ships, brewing spruce beer, gathering wood and water and food. Omai acted as interpreter as Cook heard what had happened to the men of the *Adventure:* A quarrel over theft had flared into violence. To the astonishment of the Maori, Cook did not exact *utu*—revenge. Before leaving, he liberated several pairs of the animals he had been transporting. A boar and a sow survived deep in the forests of the South Island; they begat the wild hogs that New Zealanders still call "Captain Cookers."

Contrary winds compounded the accumulated delays as Cook set his course for Tahiti in March. By this time he had meant to be well on his way to North America, ready to spend a complete summer season exploring for the Northwest Passage. Tacking against feeble easterlies, with his people out of sorts and his livestock in desperate need of fodder and water, Cook came to a bitter decision: He would not try to reach North America that year. He would stay in the tropics, and prepare for an early start northward in 1778. He was too far off course to do anything else.

The first land he spotted was Mangaia, southernmost of the Cook Islands, 617 miles from Tahiti. Ringed by a jagged reef and populated by men who brandished clubs, Mangaia offered little; and Cook sailed north to Atiu. Here his reception was better, and he is lovingly remembered to this day.

I sat on a cool veranda with the Honourable Vaine Rere Tangata Poto, a schoolteacher and Atiu's representative in the Cook Islands government. The wind rattled the fronds of coconut palms, and insects buzzed lazily in the sun, and my host told me of that episode two centuries ago.

"My ancestors had a tradition that a wonderful tribe of white people, the *Tane-mei-tai,* had once lived on Atiu, but had disappeared long since. When Captain Cook arrived, the people thought the lost tribe had returned, and they greeted them with open arms. They were happy to give food and were amazed at the power of the white men; one sailor shot a bird out of the air, and my people were a bit frightened and thought it was a miracle. They wanted the Tane-mei-tai to stay and tried to detain them by force when they started to leave; the man Omai finally convinced them that the power of the white men should not be tested." In return for the food, Cook gave one of Atiu's three chiefs a length of red cloth. "That cloth became the symbol of authority on Atiu and remained in the chief's family for decades."

Little water was available at Atiu, at Takutea, at Hervey Island; and the cattle were starving when the ships reached Palmerston Island. There, however, he found forage for the stock and abundant fish, seafowl, and greens—skillfully cooked by Omai—for the men. A classic atoll, Palmerston consists of several sandy islets enclosing a lagoon the color of sapphires. While the crewmen worked, Cook and Anderson explored.

Anderson in particular was captivated by the beauty of the lagoon and the splendor of its coral formations; for perhaps the first time, a European contemplated this enchanting realm beneath the waves. He marveled at the "number of fishes that glided gently along, playing their gambols amongst the little caverns," and the luster of their colors "far exceeding any thing art can produce." His delight moved him to a question typical of his century: "for what purpose Nature should want to conceal a work so elegant in a place where mankind could seldom have an oppurtunity by enjoying it of rendering the just praise due to her wonderfull operations."

With seven months to fill, the commander could easily have searched for other elegant works of Nature or novel communities of Man. The James Cook of the second voyage, unwilling to "Idle away" a season or a moment, would have launched a program of discovery in the South Pacific. The James Cook of the third voyage was different. He lingered for weeks on end in the Friendly Islands and the Society Islands, even when he heard that important places such as Fiji and Samoa lay just a couple of days' sailing from Tonga.

What had happened to the bold navigator, the incessantly curious man of science? Professor Beaglehole maintains that he was, simply, exhausted: body, mind, and spirit overtaxed by years of responsibility. Other authorities have suggested that he was suffering from B-complex vitamin deficiencies—caused by intestinal parasites — which can lead to distressing changes in personality. Obviously, he had reached the age when anyone might begin to hoard his strength. Moreover, this man of fierce ambition had earned a fair array of honors—and had left them barely tasted.

Whatever the cause, he was a changed man on this voyage. In Tonga he had natives flogged and even shot for petty thievery. At Moorea he burned 25 canoes and let his crew go plundering—all because of a stolen goat. Even some of his officers questioned his behavior. Of his "precipitate proceeding" on Moorea, Lt. James King wrote: "I cannot think it justifiable. . . . I doubt whether . . . punishing so many innocent people for the crimes of a few, will be ever reconcileable to any principle one can form of justice."

Both plunder and petulance figured in a plot by Tongan nobles to murder Captain Cook and his entire crew. From Palmerston Island, Cook had followed his track of 1772 to the island of Nomuka, where trade brought good supplies of food and water. A powerful young chief named Finau spoke of wealthier islands to the north; he sailed with Cook to Lifuka, primary island of the Ha'apai group. Here Cook found ever greater bounty, and he paid handsomely for it.

Thirty years later an Englishman in Tonga learned the facts Cook never knew. The wealth aboard the *Resolution* and *Discovery* enticed the local chiefs; they conspired with Finau to gain it. They planned an evening of feasting and dancing for Cook and his men. At a given signal they would fall on the Englishmen and kill them; then they would take the ships. Finau, who outranked the Lifukan nobles, thought daytime better for attacking the ships, and the other chiefs reluctantly agreed.

On the afternoon of May 20, Cook and some officers were invited to a series of dances, which greatly amused them. They were totally off guard—but the attack never came. At the last moment the Lifukan chiefs had insisted on an assault by night. Finau, incensed at this affront to his judgment, forbade any attack whatever. Cook, unaware that he owed his life to the pique of a Tongan grandee, returned to the *Resolution* for his dinner.

That night, in the hours chosen for his murder, he witnessed the most spellbinding ceremony of all his years in the South Pacific: sensuous dances performed by the women of Lifuka in the flickering light of coconut-frond torches. Naked to the waist, their breasts gleaming with scented oil, their dark hair interwoven with flowers, they swirled and undulated to the music of log drums. The Britishers were entranced, even the prudish Lt. John Williamson whose "ideas of disgust were soon changed into that of admiration. . . ."

Traditional dances still rank high in the Kingdom of Tonga. On Lifuka, I talked with jovial Fifita Tupou, acting headmistress of the Pangai primary school, as more than 500 students, aged 5 to 13, rehearsed a pageant. "We teach our songs, dances, and customs as part (Continued on page 160)

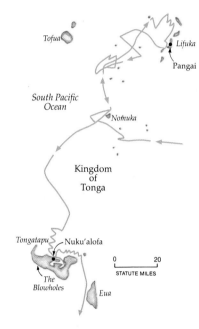

"Friendly Islands" struck Cook as the perfect name for the realm of Tonga, still an independent Polynesian kingdom and now linked with Britain in the Commonwealth of Nations. Its intricate social, political, and religious system impressed but puzzled both Cook and many later visitors.

OVERLEAF: *Women of Tonga, working with sticks and dye, decorate a ceremonial sheet of tapa, or bark-fiber cloth, as a memorial to their well-loved Queen Sālote.*

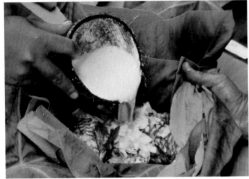

*"a baked hog eat better
than a boiled one,
that a plantain leafe made as good
a dish or plate as pewter
and that a Cocoanut shell
was as good to drink out of
as a black-jack..."*

Tonga's bounty marks a feast worthy of Cook but given for Will Gray and Gordon Gahan by the Anga 'Unga family of Lifuka. At left, Tevita Kalekale, a member of the host's string band, pares the vegetable called kape; *he wears the wraparound* tupenu *and a festive* kahoa fatai, *or lei of vines. Foods baked in an earth oven include chunks of kape wrapped in banana leaf (below, left) and lu ika, fish drenched with coconut milk and sealed in taro leaf (above). Served on raw banana leaf, the "Tongan tray" assortment (below, right) includes mullet, kape, papaya, and breadfruit. The family also offered chicken, bananas, watermelon, and lu pulu, a boiled confection of sweetened taro greens. Young pig, roasted on a spit, ranks as the prime delicacy. Mr. Anga 'Unga defines one guest's share as one pig, two pigs, "or more."*

of the curriculum," Fifita explained. "We are a small country, but our pride runs deep and our heritage is strong. Tongans have always been known as among the best dancers in Polynesia, and we want to keep that unchanged. It's not hard—every Tongan lives to sing and dance." Instead of wooden drums, empty tin cans produced the loud and lively beat; smiling broadly and clapping her hands, Fifita swayed to the music as her students danced the *kailao*, a frenetic war dance with stamping, screaming, and fierce challenges with wooden spears.

Another tradition the Tongans cherish is that of friendliness. On Lifuka, the Anga 'Unga family—three generations of it—invited Gordon and me to a Tongan feast. On a languorous afternoon near a white thread of beach, they prepared as filling and diverse a meal as we had ever encountered: suckling

Nighttime dances of thrilling precision, staged by torchlight to the thud of bamboo poles pounding the earth, hold Cook and his men enthralled at a feast on Lifuka. In fact, local chiefs had planned to catch their guests off guard, murder them, and then attack and plunder the English ships. Thirty years later another Englishman learned of the conspiracy, canceled at the last instant by its leader. Cook himself had never suspected that death was stalking him here.

pig; *lu pulu* made from coconut milk and taro leaves; fish, yams, and fruit cooked in an *umu*, or earth oven; and milk straight from the coconut. I lolled on the sand, caressed by breezes that carried the mingled scents of flowers, food, and the sea, and munched my way through course after course as the family played lilting Tongan songs.

Like us, Cook and his men left Lifuka pleasantly sated. They sailed for Tongatapu through azure waters dangerously sprinkled with coral — "a most confounded navigation," said Clerke. Cook noted philosophically: "Such resks as thise are the unavoidable Companions of the Man who goes on Discoveries." On Tongatapu the mariners reveled in festivals of dancing, kava ceremonies, extravagant feasts, complaisant women. Veterans of the second voyage might recall that in two visits Cook had stayed among the Friendly Islands only 11 days; on this journey he took 11 weeks. He became fast friends with Fatafehi Paulaho, the 36th Tu'i Tonga, the sacrosanct paramount chief. Cook described him as "the most corperate plump fellow we had met with" but also as "a Sedate sensible man" with the power of a king. "I was quite charmed with the decorum" observed in his presence, wrote Cook; "I had no where seen the like, no not even amongst more civilized nations."

In mid-August, with great relief, Cook reached Tahiti, to distribute the surviving animals sent by King Tosh to the chieftain Tu and other Tahitians of rank. He declared himself "lightened of a very heavy burden, the trouble and vexation that attended the bringing these Animals thus far is hardly to be conceived." He set his men to caulking their badly leaking ships, and persuaded them to save their usual grog ration for colder climates. He witnessed a ceremony honoring Oro, the god of war, and verified something he had long suspected—the Tahitians practiced human sacrifice. Tu urged him in vain to join in a war against Moorea. He had gained a recognized standing as "Tootee," a great chief and a friend, and with that role in Tahitian life came hints of a part he could not play.

Days brimming with festivity passed quickly. The Tahitians were awed by a fireworks display, but even more by the sight of Cook and Clerke galloping their horses along the shore of Matavai Bay. This, mused Cook, "gave them a better idea of the greatness of other Nations than all the other things put together that had been carried amongst them."

Sailing from Matavai Bay for the last time, on September 30, the captain recorded no sentiment. After that ill-starred stop at Moorea, the ships made for Huahine, where Omai had decided to settle down. Cook acquired land for him, had the ships' carpenters build him a house, and ordered some of the men to plant him some Tongan vegetables. Although Cook thought Omai a fairly shallow and silly young man, he and most of the crew appreciated "his great good Nature and docile disposition." At the time of leave-taking, according to Bayly, many tears were shed as Omai "kiss'd us & bid us a long farewell."

A final farewell to the Society Islands came on December 8, 1777. Several crewmen, comparing island life with desirable women to long duty on icy waters, had tried to jump ship. Flogging, and an impassioned speech by Cook, quelled this "spirit of Desertion." In a different spirit, Charles Clerke and William Anderson had decided to beg permission to remain in the tropics. Both were very ill with consumption, and Anderson knew that the expedition into the north could be fatal to them. Yet in the end, their plea unspoken, they sailed on with their shipmates.

On the night of December 22 they crossed the Equator—the first time that James Cook, the unequaled master of the South Pacific, had ventured into the northern part of that vast ocean. Now both Cook and Clerke expected a long and tedious haul to the coast of North America. Instead, within weeks, they came upon an unexpected paradise. A chain of high volcanic isles lured Cook toward the last of his great discoveries: the Hawaiian Islands. As he neared land, canoeloads of people approached; to his astonishment they spoke a language akin to Tahitian. They had the appearance of Polynesians, and a few soon displayed "a thievish disposition" of a familiar sort.

Capt. Charles Clerke, R.N., who sailed on all three voyages, appears in a 1776 portrait from life with a Maori warrior whose likeness the artist adapted from a drawing by Parkinson.
An Essex boy, Clerke went to sea at age 12, rising in rank by sheer merit. Courageous, openhearted, and good-humored, he became one of Cook's few intimate friends as well as a trusty subordinate officer. Fatally ill with tuberculosis, Clerke took command of the last expedition when tragedy struck Cook at Hawaii.

Cook had reached the northernmost apex of the immense triangle that enclosed the Polynesians in the Pacific. From New Zealand to Hawaii to Easter Island—some 4,700 miles in each instance — these people had colonized far-flung islands. Again he asked: "How shall we account for this Nation spreading it self so far over this Vast ocean?"

Seeking an up-to-date answer to that question, I visited archaeologist Yosihiko H. Sinoto at the Bernice P. Bishop Museum in Honolulu. "The Polynesians were uncanny sailors," he told me. "We now believe their forebears originally came from the mainland and islands of Southeast Asia. We think their direct ancestors came from around Fiji and adjacent areas, and arrived in Tonga and Samoa about 1500 B.C. The Marquesas became a major dispersal point in the east. With incredible exploits of navigation, they populated their farther islands — during maybe a thousand years. All the major migrations had ended before Columbus set sail."

I asked Dr. Sinoto what factors had impelled these great journeys.

"We don't really know," he replied with a shake of his head, "but it could have been any one—or all—of the following: rapid population increase; war; droughts; a natural desire to discover new lands. I would emphasize the last for the initial movements from the Marquesas.

"Our archaeological investigations indicate that contact continued between the closer groups, but after about A.D. 1400 the most remote points seem to have become isolated, developing distinctive cultures."

In my own journeys about the Pacific, I developed sheer awe at those migrations. In their great sailing canoes, the Polynesians had traversed endless miles of open ocean. They had taken cuttings of staple plants, shielded from the lethal salt spray. Wherever they went, they had established societies both efficient and complex.

"the swell... rolled in from the North and broke against the shore in a prodigious surf...."

Hawaii would prove the most complex of all. After anchoring in Waimea Bay on the island of Kauai, Cook went ashore—and people greeted him by falling prostrate at his feet. He was impressed by the lush stands of taro, the tidy villages, and the *heiau*, shrines with three-tiered altars. On the beach trade proceeded briskly, and with "the greatest good order."

More strictly than ever, the captain vowed to keep these new islands free of the foul "veneral" disease. He would allow no women on board, no crewmen with any sign of affliction on shore. In vain: the "oppertunities and inducements" were "too many."

Worse — though he only learned of it later — was an episode involving Lieutenant Williamson, in charge of a shore party. When excited islanders surrounded his boat, he shot a "tall handsome man" who collapsed dead in the shallows: "a Cowardly, dastardly action," said one disgusted eyewitness, "for which Captn Cook was very angry."

With that lamentable exception, three days passed in friendly trade, while the British admired the accomplishments of their hosts: beautiful bark cloth, gorgeous feather capes and helmets, as well as astonishing skill and confidence in swimming and canoe handling. Then Cook decided to reset the anchorage of the *Resolution,* but a burst of wind forced him from Waimea Bay. He tried to inch northward along the green-draped, crenellated cliffs of the Na Pali coast, but was stopped by "the force of the swell which . . . broke against the shore in a prodigious surf."

For ten more days he battled storm and current, first trying to regain Kauai, then trying to anchor off the neighboring island of Niihau. Finally, "after spending more time about these islands than was necessary," Cook pointed the prows of the *Resolution* and *Discovery* "to the Northward." Some thought the islanders "seemed to regret much our leaving . . . so soon."

Upon reflection, Cook named this group for his most consequential patron: the Sandwich Islands. As he sailed from them on February 2, 1778, he had no special comment to record, no plans for a return. Nine months later he straggled back, however, to an anchorage he would never leave.

Driftwood marks the area where Cook first came ashore, at Waimea Bay on Kauai's southwest coast, in the Hawaiian Islands —his last great discovery. Northward along the Na Pali coast (opposite), tricky currents became an exasperating menace.

OVERLEAF: *A waterfall plunges hundreds of feet down a volcanic height on Kauai.*

Chukchi
Sea

Point Barrow

Icy Cape

Arctic Circle

Cape Prince of Wales

ASIA

Cook Inlet

Prince William Sound

+ Mt. Fairweather

Bering
Strait

NORTH
AMERICA

Aleutian Islands

Gulf of
Alaska

Nootka
Sound

Unalaska
Island

0   1000

STATUTE MILES

North Pacific
Ocean

Cape
Foulweather

The Third Voyage:
"The Frozen Secrets of the Artic"

*J*n a rare interlude of clear skies, sunrise on March 7, 1778, revealed "the long looked for Coast of new Albion." Slowly Cook stood in toward a land named and first explored two centuries before by Sir Francis Drake. "There was nothing remarkable about it," concluded Cook as he recorded its hills and valleys and ever-present forests. "At the northern extreme, the land formed a point, which I called *Cape Foul Weather* from the very bad weather we soon after met with. . . ."

Very bad weather still buffets the coastline of Oregon, where Cook first sighted North America from the Pacific. On a raw June morning, I stood on the rugged headland of Cape Foulweather. Windblown rain pelted a gray swirling sea, and I braced myself against its fury. Thick mists, eddying about the rocks, quickly dampened me to the skin. Incongruous shafts of amber sunlight pierced the roiling clouds at times and scattered droplets of gold on the breakers.

After the fragrant, sparkling beaches of the tropics, this chill coastline must have dampened the spirit. "We have now very disagreeable Weather," Clerke wrote, "fresh Gales with hard squalls of Wind sleet & Snow; and a very heavy W^{t}erly Swell. It is really rather a lamentable business. . . ." The five-week passage from Hawaii to Oregon had been marked by gradually worsening weather, which at least prepared the men for deep cold. "We have been so long Inhabitants of the torrid Zone," wrote Clerke early on, "that we are all shaking with Cold here with the Thermometer at 60. I depend upon the assistance of a few good N:Westers to give us a hearty rattling and bring us to our natural feelings a little, or the Lord knows how we shall make ourselves acquainted with the frozen secrets of the Artic."

Strong gales sent the *Resolution* and *Discovery* south along the Oregon coast for five days before they could make headway to the north. On March 22 Cook named Cape Flattery, northwesternmost point of what is now the state of Washington, and that night he passed the Strait of Juan de Fuca — entrance to Puget Sound — unknowing. Rashly enough, he dismissed the possiblity "that iver any such thing exhisted." He was desperately short of water one week and several storms later, when he found a hoped-for good harbor. He had reached Nootka Sound, a jagged slice in the west coast of Vancouver Island — explored and charted by George Vancouver in 1792.

Almost immediately, the ships were surrounded by canoes. Wary at first, the Indians were assured of Cook's friendliness when he passed out servings of ship's biscuit, which they took for lovely pieces of wood. They called in welcome, "*Nootka, itchme nootka, itchme* — You go around the harbor." Thus they gained the name of Nootka. Cook found them "a mild inoffensive people," eagerly exchanging fine furs or "any thing they had" for "whatever was offered them . . . more desireous of iron than any thing else." Indeed, bright metal was so highly valued that "before we left this place," according to Cook, "hardly a bit of brass was left in the Ship, except what was in the necessary instruments. Whole Suits of cloaths were striped of every button, Bureaus &c^a of their furniture and Copper kettle, Tin canesters, Candle sticks, &c^a all went to wreck. . . ."

Obviously keen traders and excellent fishermen, the Nootkas struck the Englishmen as repulsively grubby of person. As Lieutenant King put it, "their faces were bedaub'd with red & black Paint & Grease . . . as their fancies led them; their hair was clott'd also with dirt, & to make themselves either fine, or frightful, many put on their hair the down of young birds, or platted it in sea weed or thin strips of bark dyed red. . . ."

"But," Bayly added, "even thro all this dirt & nastiness the fine rosy

PRECEDING PAGES: *Chill —and danger —mark Cape Foulweather. Seeking the fabled Northwest Passage, Cook made landfall here; storms quickly repelled him.*

bloom of youth appeared on the Cheakes of many of them." Some of the young gentlemen, reported surgeon's mate David Samwell, devised a "Ceremony of Purification"—a warm soapy bath—for young women. After taking pains with their "red oaker" makeup, the girls were astonished to find the Englishmen scrubbing it off. "Such," concluded Samwell, "are the different Ideas formed by different nations of Beauty & cleanliness...."

Nowhere among "uncivilized nations," said Cook, had he found stricter ideas of property rights than among the Nootka, who wanted payment for water, wood, and grass. When he questioned this practice in a stern quarter-deck tone, one man caught his arm, gave him a push, and gestured for him to be off. Cpl. John Ledyard of the marines, a native of Connecticut, recalled that Cook smiled and paid, exclaiming, "This is an American indeed!"

I talked with anthropologist Susan J. Anderson of British Columbia's Provincial Indian Programs, about the Nootka and neighboring tribes. "These coastal people were unique," she said. "They had no crops to plant, no livestock; they relied on hunting, fishing, and gathering food. But their environment was so rich, and they used it so skillfully, that they developed a vigorous economy and strong traditions of art. They hunted seals, sea lions, and whales. They moved to the shores in spring for seafood; they took the different kinds of fish and berries and roots in season; and they moved up to the heads of the inlets in the fall. During the winter they gathered by clans in large planked houses and passed the cold months with a ceremonial life of great majesty and diversity."

Cook and his men saw little of the rituals, but noted some important ceremonial objects, deftly carved masks and large wooden images and capes of wolf skin. They often enjoyed Nootka music; King mentioned "harmony that equally surprized & pleased us," and Burney "strange placed Notes, all in Unison and well in tune."

A short stop for water evolved into four weeks when the carpenters discovered rot in the *Resolution*'s masts. Trees from the copious forests of spruce, hemlock, and cedar had to be chosen, felled, trimmed, sawed, and fitted — time-consuming work, but a life-or-death matter. Cook, in the meantime, assessed the local greens—nettles and wild garlic—and scouted new shores. A crew of midshipmen rowed him about Nootka Sound, a steep-walled, green, and chilly bay that reminded me of Dusky Sound.

James Trevenen, one of the oarsmen, later recalled: "We were fond of such excursions, altho' the labour of them was very great, as ... more agreeable than the humdrum routine on board...." They "gave us an opportunity of viewing the different people & countries, and ... we were sure of having plenty to eat & drink, which was not always the case on board.... Capt. Cooke also on these occasions, would sometimes relax from his almost constant severity of disposition, & ... converse familiarly with us. But ... as soon as we entered the ships, he became again the despot."

His ships repaired once more, Cook sailed on April 26; the Indians, waving their new saws and hatchets and other tools, sang "a parting song" while a dancer performed in various masks. It was in fact a farewell blessing, and the animal masks evoked spirit guardians to strengthen the song; but, as Cook said, he had not gained "the least insight into their Religion."

Almost immediately a storm struck, rising to "a perfect hurricane" and driving the ships far out to sea. Land was not sighted again until May 1, off what today is southeastern Alaska. Clouds and snow showers vanished, and Cook named Cape Fairweather and Mount Fairweather in celebration.

He followed the northwest trend of the coast as quickly as the gentle breezes permitted. The Admiralty had ordered him "not to lose any time in exploring" or anything else while hastening to latitude 65° N., just below the Arctic Circle, there to begin his search for the Northwest Passage

"At the
northern extreme,
the land
formed a point,
which I called
Cape Foul Weather
from the
very bad weather
we soon after
met with...."

Yaquina Head Lighthouse beams a warning to mariners near Cape Foulweather on the Oregon coast. In March of 1778, Cook worked without beacons—and with utterly inadequate maps—tacking off a coast glimpsed only at intervals. Unable even to trace the shoreline, he struggled north for days, his ships dancing about with the harsh whims of weather, before locating a safe harbor. Skills hard won in northern waters served him well here. Though sailing conditions might evoke the Yorkshire coast of his youth, the tide pools shelter a Pacific fauna—including the giant green sea anemone and the California mussels—found in no other waters.

in June. At the head of the Gulf of Alaska, however, he entered an inlet heading due north, hoping it would be part of the coastline. Gore, thinking it might be the long-sought passage, named a nearby point Cape Hold With Hope. Both were disappointed: Prince William Sound proved a cul-de-sac. Sailing out ten days later, Gore named another point Cape Lost Hope.

Within days the expedition came upon another deep-water channel between towering snowclad peaks. Gore fairly peppered the land with optimistic names: Hopes Return, Mount Welcome, Land of Good Prospect. He and others insisted on exploring this waterway. Cook "was fully persuaded that we should find no passage," but deferred to his officers.

For 11 days the ships plied the waters of the stunningly beautiful route now called Cook Inlet—to another dead end. Where the city of Anchorage stands today, the inlet divides into branches that could not possibly lead to Hudson or Baffin Bay. "If the discovery of this River [the inlet] should prove of use, either to the present or future ages," Cook rather sourly wrote, "the time spent in exploring it ought to be the less regreted, but to us who had a much greater object in View it was an essential loss; the season was advancing apace. . . ." Gore, for once, remained silent.

Cook's orders gave him that year in which to succeed or fail. Then he was to return to England by whatever route seemed best. He was expected home in 1779. In Paris that spring, Dr. Benjamin Franklin—winner of the Copley Medal in 1753, Fellow of the Royal Society since 1756, and now diplomat of the United States—issued a letter for all American warships: Captain Cook should not be considered an enemy, attacked, taken prisoner, or delayed. American commanders should treat him "with all Civility and Kindness" and give him any possible assistance he might need. The work of "that most celebrated Navigator and Discoverer" would serve not only the sciences but also trade and commerce and "the common Enjoyments of human life."

Enjoyments were scant as Cook groped along the Alaska Peninsula through thick mists and chill rains. Clerke cursed "the confounded fog" and foul wind. The peninsula was forcing the ships southwestward for 725 miles, to about 54° N.; and then in the Aleutian chain they almost met disaster. Dense fog ensnared them off Sedanka Island and the larger Unalaska. Even in daylight, wrote Cook, "we could not see a hundred yards before us, but as the wind was now very moderate I ventured to run. At half past 4 we were alarmed at hearing the Sound of breakers on our larboard bow. . . ." The ships had run among rocks that protrude from the foaming gray water like the fangs of a mythical sea beast. Cook shouted for the anchors. "A few hours after, the fog cleared away a little and it was percieved we had scaped very emminant danger. . . . Providence had conducted us through between these rocks where I should not have ventured in a clear day. . . ." Clerke wryly observed: "very nice pilotage, considering our perfect Ignorance of our situation."

Free to turn north at last, Cook prowled through the mists of the Bering Sea. He had two sketchy maps that supposedly showed the findings of Vitus Bering and Alexei Chirikov fifty years earlier. As he plodded north across open water, one gave a maze of islands and the other solid land.

The ships were bucking "a nasty jumbling Sea" on August 3 when William Anderson finally succumbed to his long disease. Everyone mourned him. "He was a Sensible Young Man, an agreeable companion," wrote Cook, stressing the varied knowledge "that had it pleased God to have spar'd his life might have been usefull in the Course of the Voyage."

King wrote of the surgeon's fortitude—knowing himself doomed, he had remained self-controlled and serene. With due ceremony Anderson's body was committed to the deep. "If we except our Commander," said King, "he is the greatest publick loss the Voyage could have sustaind."

Sunlight on August 9 gave Cook a clear look at the "Western extremity of all America hitherto known." He named it Cape Prince of Wales, and it marks the eastern shore of Bering Strait at its narrowest, between North America and Asia. Just 54 miles away sprawls Siberia, with only Little and Big Diomede Islands in between. At the cape, Cook and his men caught a glimpse of some natives; the descendants of those Eskimos still live here. In a small twin-engine airplane, Gordon and I flew in to the huddled village of Wales, hoping to get to Little Diomede, where 130 Eskimos still follow a fairly

traditional life. Our luck was with us.

June's midnight sun still perched on the horizon when we spotted an open boat coming in: a rare survivor from the Eskimo past, a walrus-hide umiak. Nearly 35 feet long and 6 wide, it had several tanned and intricately stitched hides stretched over the handhewn wooden framework.

"My father built the frame of this boat," John Iyapana told me as it was pulled onto the gravelly shore. "Some of the wood in there is older than me. But we have to change the hides every season." A wiry, compact man, John has the high cheekbones and broad face typical of his people; his sharp black eyes reveal a wily intelligence. Captain of this boat and an elder of Diomede, he and his ten-man crew had called at Wales for mail and supplies after a successful walrus hunt; he agreed to let us accompany him home.

On a bright and cloudless day we pushed the umiak into the frigid gray surf. Soon we were dodging among ice floes that stretched as far north as I could see. They ranged in size from ice cubes to hundreds of feet across. On several of the larger ones were huge walruses—as many as a hundred to a floe—gigantic, blubbery mammals with comical faces and long ivory tusks. They grunted, trumpeted, bellowed, and belched as we neared them. If we came too close, they would heave themselves across the ice and plunge into the water with tsunami-like splashes.

"My people have always relied on walruses for so much," said John.

Hunters and traders, Nootkas protect their families and goods from the cool, humid climate in log-and-plank houses. On Vancouver Island at Cook's first anchorage, these Indians swapped everything from grass to sea-otter pelts for bits of metal—and unwittingly set off a fur-trading boom. John Webber, who portrayed the armed man opposite, bartered brass buttons from his jacket for permission to draw the room's pillarlike carvings.

OVERLEAF: *This Cook Inlet arm offers no passage to the north. Recognizing the fact, Cook logged a "triffling point in Geography" aptly named Turnagain.*

"They give us meat and oil, hides, and ivory for tools and carving. Ivory carving is now the main source of income for our village. We carve all winter while it is dark, and sell our work in Nome and Fairbanks and Anchorage."

During four stormy days in the clustered village of Ignaluk, Gordon and I visited the homes of several craftsmen. I watched Sam Mogg dry and carefully prepare pieces of ivory, Dwight Milligrock carve lifelike figurines of Eskimos and the animals they hunt, and Albert Iyahuk use a small mouth drill turned by a hand bow to bore holes in beads he had patiently shaped.

One afternoon when the gales abated, I clambered 1,300 feet to the top of Little Diomede. The island is just a pile of weathered boulders, interspersed with mosses and grasses. As I scrambled up the rocks I disturbed hundreds of the million-odd birds that nest here during the summer.

At times the sky seemed blackened by the swirling shapes of auklets, puffins, murres, and gulls; their shrill cries chased me to the summit. Balanced on a final rock, I could see the misty hump of Cape Prince of Wales in the east, flat-topped Big Diomede to the west, and the ragged coast of Siberia beyond. The ice, pushed by rampaging winds, had disappeared far to the north.

"very nice pilotage, considering our perfect Ignorance of our situation."

Like Captain Cook before us, Gordon and I followed the ice. On August 17, 1778, Cook "percieved a brigh[t]-ness in the Northern horizon like that reflected from ice, commonly called the blink. . . ." Four degrees north of the Arctic Circle, the *Resolution* and *Discovery* confronted Arctic ice for the first time: "quite impenetrable . . . from WBS to EBN as far as the eye could reach." The ships dropped a few miles south; in shoal water Cook saw a spit of land "much incumbered with ice" and duly named Icy Cape. Wind crowding him toward a lee shore brought "the main body of the ice . . . driving down upon us." Quick tacking and a shift in the wind rescued him once again.

Far from the nearest human community, Icy Cape today is as forbidding as when Cook named it. I walked along its rocky strand about 2 a.m. The sun still glimmered in the sky and shone glancingly off the frozen ocean. The gray-green mounds of ice, piled in convoluted jumbles on the beach, imparted a numbing chill that penetrated every layer of clothing. I scanned the shore for treasures and noticed many small ones: bird feathers, colorful rocks, delicate shells. Then treasures of a different magnitude: the bleached skeleton of a whale, the moldering carcasses of two walruses, the paw-prints of a polar bear. As I scrutinized these, the wind brought the faraway yapping of an arctic fox.

"Confounded fog" lifts to unveil jagged rocks off Sedanka Island. The expedition lost 16 days of good sailing in Cook Inlet, and forfeited 7 degrees latitude and 15 degrees longitude in the subsequent search through drizzle and rain for a break in the southwest-trending Alaska Peninsula. The smile of fortune seemed reserved for rescue from close calls in this otherwise-luckless voyage to the Arctic. Running in a thick mist off Sedanka, Cook heard breakers and immediately dropped anchor. Relenting fog, hours later, revealed the ships' blind passage "between these rocks where I should not have ventured in a clear day. . . ."

On the large ice floes off the cape, Cook's men saw hundreds of walruses—sea horses, they called them. Cook sent boat crews to shoot a few for fresh meat. Always experimental in diet, he reported that the "fat at first is as sweet as Marrow . . . the lean is coarse, black and rather a strong taste, the heart is nearly as well tasted as that of a bullock." Trevenen commented: "Captain Cook here speaks entirely from his own taste which was, surely, the coarsest that ever mortal was endured with." Many found that walrus meat "produced purgings & vomitings."

On this controversial ration Cook zigzagged about the Chukchi Sea for ten days, seeking a route through the ice. In drizzling rain he closed with the coast of northeastern Siberia. Always the ice defeated him. On August 29 he wrote: "I did not think it consistant with prudence to make any farther attempts to find a passage this year in any direction so little was the prospect of succeeding." Lieutenant King expressed a veteran's relief: "Those who have been amongst Ice, in the dread of being enclosed in it, & in so late a season's can be the best judge of the general joy that this news gave."

Indeed, it was a wise decision. During some summers, the pack ice of the Arctic Ocean does recede far enough to permit sailing along the tricky northern fringe of Alaska and Canada. Luckily 1778 was not one of those years. If he had been able to round Point Barrow and pick his way east through the shoals, he would have been trapped eventually. Prevailing northerlies bring the pack down in September, while young ice forms inshore — sharp enough when it thickens to gash a wooden hull as coral might do. Cook's ships, for the sake of speed, had not been given special sheathing for Arctic waters; even with it, they could easily have been ground to splinters. He and his men would have been forced to bivouac as best they could. Anyone who did not freeze or starve during the winter could attempt to stumble back to civilization when the cold relented in May or June. And the nearest English outpost, at Fort Churchill on Hudson Bay, was more than 2,000 unmapped hopeless miles to the southeast.

Religion survives where Russian fur traders once sojourned. A descendant of the original inhabitants, the Aleuts, the Reverend Ishmael Gromoff lights candles in the Russian Orthodox Church on Unalaska; one of its icons gleams above. Russian influence in the Aleutians began three decades before Cook's visit. With makeshift communication and a supply of rum, he and the Russians shared maps of the region and speculated on coastlines.

A sudden blustery snowstorm on the night of August 29 heightened Cook's awareness that the season was "very far advanced." Ice filmed the ships' drinking water as they ranged south near the coast of Siberia, the leadsmen taking frequent soundings. Cook's maps were proving almost as foggy as the weather. Thinking that he could clear up some confusing points about the American coast, he "steered over" for it on September 5.

On October 3, he anchored at Unalaska in a harbor he had found in July. Water was plentiful, with wild berries still in season, halibut and salmon abundant and some of the latter "in high perfection." The carpenters could repair the leaking seams on the *Resolution*'s starboard side. And Cook and the officers had the delightful surprise of meeting a European gentleman in the wilds of Unalaska.

For three decades the Russians had slowly been expanding their fur trade throughout the Bering Sea, and Unalaska had become one of their headquarters. They traded tobacco and trinkets to the local Aleut population for the luxuriant and valuable pelts of the sea otter. When Cook reached the area, the English had no idea of its potential; but his men learned of its wealth on this voyage.

Some who had acquired sea-otter pelts from the Nootka eventually sold them to Chinese merchants at a good price. Trevenen had exchanged a broken brass buckle for a prime skin that brought 300 Spanish dollars. William Bligh had given some Tahitians a shilling hatchet for 30 large green glass Spanish beads; at Prince William Sound he gave 12 beads for 6 pelts; and in China he got the equivalent of £15 for each fur—a gross profit of 449,900 percent.

As Dr. Franklin foresaw, Cook's expedition proved a keen stimulus to a new pattern of trade. By the early 20th century the sea otter and other mammals of the North Pacific had been hunted nearly to extinction. Only a nick-of-time international treaty, signed in 1911, saved the sleek sea otters and fur seals that once thronged the cold shores of this ocean.

In the small Russian community on Unalaska, the dominant figure was Gerassim Gregoriev Ismailov, a man *(Continued on page 184)*

Before migrating south, auklets provide delectable meals for those swift with a net during the summer nesting. Each year Eskimos of Little Diomede Island scour the rookeries for the birds and their eggs. In search of larger prey, the men of a hunting party set out in an umiak, a boat of walrus skin and driftwood. The bounty of this season in the Arctic offered no consolation to Cook's cold and tired crew; Clerke proclaimed the region "a damn'd unhappy part of the World."

"On the ice lay
a prodigious
            number of
Sea horses and
      as we were
in want of fresh
   provisions the
boats from each ship
      were sent to
         get some."

With walrus in sight, rifle in hand, the umiak leader signals alert in the Bering Strait. His targets, male walruses, flash their pickax-like tusks as they slide off an ice floe and out of sight. Raising the head and displaying the tusks, as if in threat, marks their action as hurried. John Webber depicted them under fire by Cook's boatmen in August 1778. The walrus herd encountered in the icier waters above the strait struggled to escape at the first gunshot. Described as "marine beef," but not digested so readily, the walrus catch in the Arctic phase of this voyage did supply variety in the crews' diet, rigging for the ships, and oil for the lamps.

of education and experience. He welcomed the visiting captains by sending them each a pie with a rye crust, "well raised and light," and a filling of salmon, "highly seasoned with peper." He and Cook shared dinners, ship's wine and porter and rum, and geographical information. Unfortunately they had no common language. Cook "felt no small Mortification in not being able to converse with him any other way then by signs assisted by figures and other Characters which however was a very great help." No one in Cook's party, and none of the Russian "Seamen or Furriers," could do better. Ismailov, however, pointed out errors in the old maps Cook had, and added considerable information of his own. He even produced two manuscript charts for Cook to copy. One showed "the coast of *Tartary*" and the Kurile Islands. The other—"to me the most intresting," Cook said—showed

"all the discoveries made by the Russians to the Eastward of Kamtschatka towards America." Except for the contributions of Bering and Chirikov, Cook concluded, these findings would "amount to little or nothing." He must have thought with satisfaction that much remained for him to clear up —and claim for the Crown.

All in all, Cook was gratified by the encounter with Ismailov and his Russian comrades, the first Europeans he had met in his Pacific explorations. Ismailov graciously took into his care a letter from Cook to the Admiralty, and one to his wife, to be forwarded to Kamchatka in the spring of 1779 and thence to St. Petersburg by winter, possibly to reach London in 1780.

Although the Russians have long since departed, their influence still lingers in Unalaska, particularly in religion. I learned of it with the Reverend Ishmael Gromoff, himself a native Aleut, pastor of the Russian Orthodox church on the island. A large white structure topped by the traditional onion-shaped domes, it dominates the small frame houses comprising the Aleut village of Unalaska, and today it ranks as the oldest in America still serving a congregation of that faith.

"The church itself dates back more than a century and a half, to 1825. But the first Russians arrived here in the 1760s," Father Ishmael told me as he guided me through the elegantly furnished building. Spacious and high-ceilinged, it exuded the mingled scents of aged wood and musky incense. Rich icons and expressive oil paintings adorned the walls, and brass stands throughout the sanctuary held hundreds of candles.

"The earliest merchants baptized the Aleuts," the priest continued, "and with the coming of the priests our religion quickly spread through the area that became known as Russian America. One of the early priests who came to Unalaska spoke of the 'enthusiasm, nay, craving for God's Word' among the Aleuts. As the religion strengthened, churches were needed and soon they were built: in Kodiak, in Sitka, and then here. These places remain the strongholds of our faith in this region. Even though many changes have come to Unalaska in the last few decades, the Russian Orthodox religion remains strong, and we still have an active congregation."

Even during Cook's visit, the religion was evident. Samwell, candid as usual about his interest in women, noted that the Russians "always expressed their disapprobation of our intercourse with the Indian Women, & with a very grave Phyz seemed to lament our depravity in having Connection with those who they said were 'neet Christiãne', that is not Christians. . . ."

Cook brought out his Eskimo wordlist, which the Admiralty had provided for possible use along the Northwest Passage, and did his best to collect words akin to those of Greenland. He had discovered that different languages were spoken along the American shores, but suspected "that all these nations are of the same extraction. . . ." For once he may have yielded to a little wishful reasoning, because he thought this implied "communication of some sort by sea between this Ocean and Baffins bay." Hastily he added a note that this route "may be effectually shut up against Shipping, by ice and other impediments."

Winter was advancing rapidly upon Unalaska. Cook, his men refreshed with good provisions and his ships restocked, had made his plans for the coming year. He would make for the tropics now; and next spring he would return north, to Kamchatka, "endeavouring to be there by the Middle of May next." His eagerness for new discoveries, apparently slack in his stay at Tonga, seemed to have returned. In September he had written of the "great dislike I had to lay inactive for Six or Seven Months, which must have been the case had I wintered in any of these Northern parts. No place was so conveniently within our reach . . . as *Sandwich Islands*, to these islands, therefore, I intended to proceed. . . ." He had seen little of them, after all. He would miss Anderson, whose grasp of the language and whose sensitive perception would have helped fulfill the duty outlined in the Admiralty's instructions: "to observe the Genius, Temper, Disposition, and Number of the Natives . . . shewing them every kind of Civility and Regard; but taking care nevertheless not to suffer yourself to be surprized by them, but to be always on your guard against any Accidents."

Catch becomes craft when Dwight Milligrock carves walrus ivory into Eskimo genre scenes, like the one in his hand. The traditional winter pastime for Little Diomede has become its economic mainstay. Children of Eskimo lands inherit the task of preserving their culture, rich long before Cook arrived, and adapting it to a modern world.

Unwittingly, Cook had plotted a fatal course. The wind veered and heightened to a gale when he left Unalaska on October 26. Shoddy rigging claimed a victim on the *Discovery*. As a midshipman noted tersely, "In a very heavy gust of wind, the fore & Main tacks gave way, which kill'd Jnº MᶜIntosh Seaman & very much hurt the Boatsʷⁿ & others."

No doubt the commander resolved to discuss his ships' defects when he returned to report to their Lordships. Young midshipman Trevenen analyzed him well: "indefatigability was a leading feature of his Character. If he failed in, or could no longer pursue, his first great object, he immediately began to consider how he might be most useful in prosecuting some inferior one. . . . Action was life to him & repose a sort of death."

Lonely Eskimo hunter trudges home at Point Hope with food for the family. Using the unaak *in his hands, he can test for weak ice and retrieve ringed seals he has shot. If Eskimos of Cook's time could have shared their knowledge of land, sea, and seasons, the captain might have unlocked the Arctic secrets that defeated him. North from here, Cook sailed nearly to Asia searching desperately for a route through the southward-moving ice pack.*

OVERLEAF: *Scattered ice floes mirror shattered hopes at Icy Cape, the northernmost landfall in Cook's voyage. The crunch of the ice and the freeze in the air sent Cook in retreat, turning his thoughts to necessities of wood, water, and wintering. Reluctant to leave, determined to return, Cook yielded to that August ice pack's message — that time was running out.*

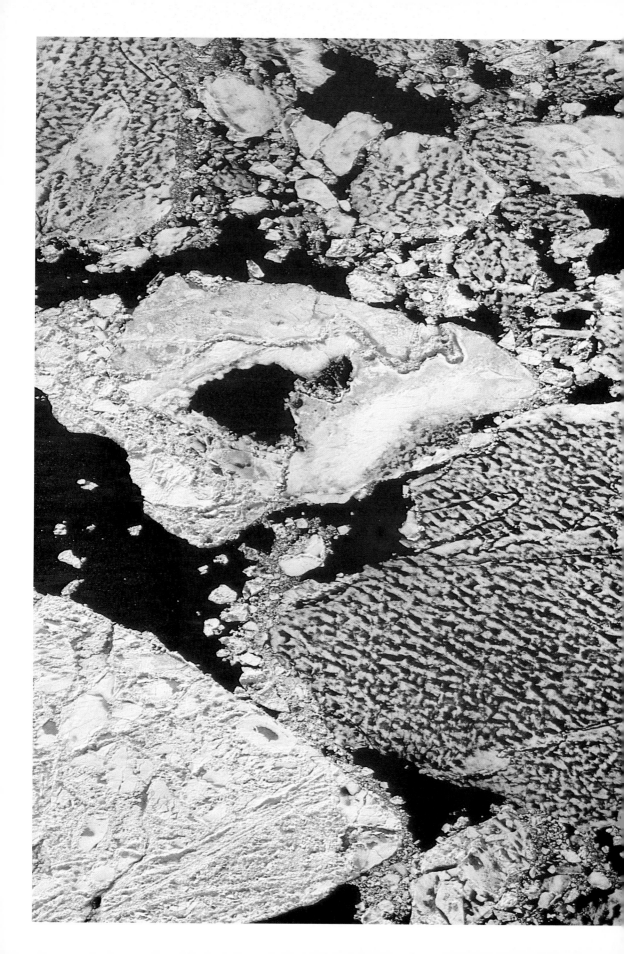

The Third Voyage:
"The Man They Worshipped as a God"

ilting breezes from the tropics soon replaced the icy gales of the Arctic as the *Resolution* and *Discovery* steadily cleaved the waves toward Hawaii. In less than thirty days the temperature had risen more than forty degrees, and Charles Clerke, aware of his precarious health, welcomed the change: "we are now advancing towards a Climate, which we have all reason to suppose we shall find less turbulent & boisterous, than those we have just bid adieu to. . . ."

At dawn on November 26, a smudge of land appeared on the southern horizon. Soon Cook could discern "an elevated saddle hill whose summit appear'd above the Clouds, from this hill the land fell in a gentle sloap and terminated in a steep rocky coast against which the sea broke in a dreadfull surf." He had fallen in with the Hawaiian Islands from the east, and the first he spotted was Maui, dominated by a broad-shouldered volcano 10,000 feet high. As the ships closed the coast, the men could see cultivated fields, houses, and people—but Cook made no move to anchor.

Indeed, for all of fifty days he coasted the islands of Maui and Hawaii and never attempted to land. Always in the past he had sought out a harbor almost immediately, not only for fresh provisions but also to let his people escape the closetlike constraints of a small ship. But now things were different. Officers and men became increasingly puzzled and frustrated. For months they had endured the frigid dark barrenness of the far north and now, with paradise in sight, they were confined on board.

Cook remained adamant in his new policy, which he did not explain. Upon sighting land he had issued orders flatly forbidding private trade, unauthorized use of firearms, and shipboard trysts with island women. To his horror he found the "Venereal distemper" afflicting the very first islanders he met off Maui—apparently transmitted south from Kauai since his visit early in the year. These friendly and appealing people eagerly brought canoeloads of food out to the ships, and crowded on board to trade fish, pigs, fruit, and roots for nails and iron implements. According to Midshipman Edward Riou, the women eagerly indicated that they would "grant favors to any one who would be so kind as to ask them In, but the orders . . . necessitated the poor ladies to go on shore." Indignant, the women abused the sailors with rude gestures and unmistakable raucous insults.

After several days of this, Cook took advantage of the wind to make south to the island of Hawaii, surmounted by the massive, snowcapped volcanoes of Mauna Loa and Mauna Kea. He continued to keep the sea, going inshore to trade during the day and standing offshore at night. For a fortnight no one set foot on this new island, and resentment was burgeoning among the crew. For no apparent reason they were still on short rations of food, and something like a crisis blew up over drink.

The first fresh provisions had included sugarcane, and Cook had it worked into "a very palatable and wholsome beer." He and his officers liked it; he wrote that it "was esteemed by every man on board." He ordered it served instead of the normal grog ration. It seems he was misreading his own men; he soon learned that "not one of my Mutinous crew would even so much as taste it." The sailors gave him a letter complaining of the "beer," which they considered unhealthful, and the food situation. Cook was incensed, according to Midshipman John Watts; he "address'd ye Ships Company, telling them He look'd upon their Letter as a very mutinous Proceeding & that in future they might not expect the least indulgence from him."

What had happened to James Cook, the kindly but firm father figure of the first two voyages? How could the captain who had so carefully nurtured

PRECEDING PAGES: *In dusky sunlight a lone Hawaiian outrigger slices through placid seas. For weeks Cook explored these waters—and then tragedy struck.*

the respect of his men now dismiss them as mutinous rabble? Where was the even-handed administrator of justice — who, though quick to anger, was always coolheaded in judgment? No one really knows. Cook was keeping his own counsel, not discussing plans with his most trusted officers.

Soon after losing his temper in this episode he almost lost his ship.

After a nasty storm on the night of December 18 the sea carried him toward land; under a full press of sail he had a fighting chance of safety. Rigging broke and three sails split at the most awkward moment; he escaped by the narrowest of margins. Cook sat down with his journal and penned— in measured, formal terms — a bitter indictment of the Navy Board for supplying gear of inferior quality. This implicated his good friend and benefactor Sir Hugh Palliser, and his great patron Lord Sandwich, the officials responsible. He might well seem distracted if he had come to suspect that—by corrupt dealings or by over-sight—his superiors had played him false.

As he tried to round the eastern cape of Hawaii, the ships met unfavorable and unreliable winds as well as strong adverse currents. The *Resolution* and *Discovery* be-came separated and did not find each other for two weeks, aggravating his worries.

Finally, his men frazzled and his ships leaking and tattered, Cook was forced to search for an anchorage. Half-way up the west coast of Hawaii, Master William Bligh found a dent in the lava-draped coastline. Entering Kealakekua Bay, the ships were escorted by "an immense Fleet of Canoes, near 800," noted Burney. The Hawaiians greeted the Englishmen with curiosity, cordiality, and en-thusiasm; "nor was the Pleasure less on our side," King observed, "we were jaded & very heartily tir'd, with Cruis-ing off these Islands for near two months. . . ."

With the prospects of solid land under their feet, ample food under their belts, and ardent lovers in their arms, the crewmen soon began to forget the disappointments and dangers and frustrations of the previous weeks. And Cook quickly became so immersed in island affairs that he could not even maintain his journal.

His last entry, dated January 17, 1779, recorded an overwhelming welcome: "I have no where in this Sea seen such a number of people assembled at one place, besides those in the Canoes all the Shore of the bay was covered with people and hundreds were swimming about the Ships like shoals of fish." So many Hawaiians clambered aboard the *Discovery* that she heeled sharply. Finally in all the hub-bub Cook found two chieftains who cleared the decks; they introduced a wizened old priest named Koa who approached the captain with an offering of two coconuts and a small pig. Koa draped Cook with a piece of red cloth, and repeated a long incantation.

That afternoon Cook went ashore. He landed on a small beach tucked beneath the precipice, the *pali*, that looms above the bay. Ritual engulfed him. Except for priests, all the Hawaiians prostrated themselves before him. In a dignified procession, Koa guided the captain and his party to a nearby *heiau*, a large tiered platform of stone. As they walked, the Englishmen heard one word repeated again and again: "*Erono, Erono*." Hauntingly, it echoed from the cliffs. Cook entered the sacred area, which was bedecked with human skulls and images carved in wood, six feet tall. Before one statue stood an altar that held a decaying hog. After a prayer at

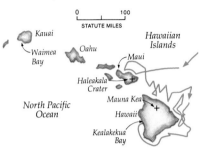

Escaping the ice-clad Arctic, the Resolution *and* Discovery *cruised south to the Hawaiian Islands for the winter of 1778-79. The lush shores of Maui finally hove into view. Cook's men anticipated the pleasures of a tropical climate, abundant food, and compliant women. The captain, for unstated reasons, dashed their hopes: He refused to land. For fifty days, the ships coasted Maui and Hawaii, on meager rations of fresh food. The crew angrily confronted him. Cook, debilitated by the strain of years of exploration and possibly wracked by disease, cursed his "Mutinous crew," but agreed to anchor.*

this spot, Koa escorted the captain onto a rickety scaffolding. There he wrapped Cook in another ceremonial sash of red; and with a young priest named Keli'ikea he voiced a kind of litany. Koa led Cook down to make a circuit of the images, and seemed to berate most of them; but at one, half the size of the others and swathed in cloth, Koa "prostrated himself, & afterwards kiss'd [it], & desir'd the Captⁿ to do the same," as King reported. Cook "was quite passive, & sufferd Koah to do with him as he chose."

Cook's men were led to the central area where the captain was seated between two images. A procession of islanders brought them offerings of food, and Keli'ikea uttered "speeches or prayers . . . answerd by the Croud repeating the word Erono." At last Cook rose, distributed a few trinkets and gifts of iron "for the Eatooa"—that is, for the god—and all the strangers were

escorted from the heiau with great solemnity. King recorded that "two men with wands went before the Captain repeating the same word as before, & on which all prostrated themselves. . . ."

This astonishing pageant left the Englishmen confounded. King summed it up as a "long, & rather tiresome ceremony, of which we could only guess at its Object & Meaning, only that it was highly respectful. . . ." Cook could never know how his arrival meshed with religious beliefs. As he was always quick to emphasize, these are obscure to outsiders; but an array of evidence suggests that in Cook the Hawaiians saw the incarnation of Lono— "Erono" to English ears — their benevolent god of peace and light and abundance. Lono's symbol was a wooden staff with a square banner of tapa cloth secured to a crossbar; the masts and spars and sails of the *Resolution* and *Discovery* were in keeping. Cook showed dignity and generosity, attributes of Lono, and he arrived in the midst of *makahiki,* the season that ran from October or November into February. This season combined a great gathering of tribute, for the rulers, with dances and games and feasts for all. Cook anchored with festivities in full swing. Finally, legend foretold the return of Lono to his people, bringing gifts.

Ritual honor surrounded Cook whenever he went ashore, and *tapu* protected a field assigned to his party. Great pomp marked the arrival of the island's foremost *ali'i,* whom the English called the king. Hundreds of canoes attended this dignitary as his craft brought him into the bay. Kalei'opu'u was an amiable old man, affected by years of kava drinking—he was scabrous, emaciated, and palsied—but still dignified in bearing. In him, it seemed, secular power merged with sacred; and he acknowledged Cook with extraordinary honors. In a rite of exalted formality he greeted Cook, removed his own exquisite cape of birds' feathers in the precious hues of gold and the sacred red, and draped it around Cook's shoulders. He presented a feather-trimmed helmet and other regalia of high rank. Still compliant and passive, the captain offered gifts in return—ironwork that the ships' armorers hastily prepared with fire and tongs and anvil.

"A constant exchange of good offices, & mutual little acts of friendship obtained among us." So Trevenen summed up long happy days of peace

Cliff-lined Kealakekua Bay bites into the lava-draped west coast of Hawaii. The Resolution's *master William Bligh—later captain on the* Bounty—*found the bay, the best refuge on this coast. As Cook's ships glided in, hundreds of canoes and thousands of Hawaiians engulfed them. "I have no where in this Sea seen such a number of people assembled at one place," wrote Cook in amazement. Soon Kalei'opu'u, the island's highest-ranking chief, arrived in an elaborate double-hulled sailing canoe (left) to welcome Cook with pomp worthy of a king and to offer him tribute.*

with the islanders. The educated visitors recorded their impressions in detail, including the first admiring accounts of Hawaiian surfing; the carpenters carried on their repairs; and the men, "mutinous" only a month before, now seemed ecstatic. "We live now in the greatest Luxury," remarked Samwell, "and as to the Choice & number of fine women there is hardly one among us that may not vie with the grand Turk himself." Inevitably, Cook had been forced to relax his rule concerning women, and soon they overran the ships. Even so, ritual respect held pilfering to a minimum.

The veneration accorded Cook extended in lesser degree to King, revered as Cook's son, and to Clerke, who would have none of it. He asked the chiefs to forbid prostrations in his honor because "I disliked exceedingly putting so many people to such a confounded inconvenience."

Even as exotic mortals, the first European visitors inevitably disrupted the fabric of life they found. Offerings worthy of a god further taxed the supplies of food on Hawaii. Kalei'opu'u and the chiefs, wrote King on February 2, "became inquisitive as to the time of our departing & seemd well pleas'd that it was to be soon." Cook was planning to explore the rest of the Hawaiian Islands—and then go north for another summer in the Arctic.

Amid fanfare and general good feeling, the *Resolution* and *Discovery* sailed early on February 4, seen off by a throng of canoes. They worked slowly northward in unsettled weather, *(Continued on page 202)*

"the inland country rises...abruptly to a mountain, which
is broken at the top, which must be very high, since we
think we can discern a good deal of snow upon it...."

196

*R*umpled summit of Mauna Kea — at 13,796 feet the tallest peak in the state —
soars toward clouds banked above Hawaii. Between surf and snow, this large
island has more varied natural resources than any other in the Hawaiian group.
Yet even so, the presence of Cook's men strained the supplies of food available.

*C*leaving the breakers (below), a canoe team
launches an outrigger in the surf of Kauai. At right,
the steersman prepares to clamber aboard for an
11-mile, open-ocean canoe race. "These People handle
their Boats with great dexterity," noted Clerke; he
called all their canoes "exceeding good."
Such craft brought food and women to Cook's men—
and soon sated the once-restive crew. "We live now in
the greatest luxury," sighed one young man.
The luxuries came not simply from keen trading, as
on other islands, but from a misunderstanding: The
Hawaiians perceived James Cook as one of their gods.

"These People handle their Boats with
great dexterity, and both Men and Women are
so perfectly masters of themselves in the Water,
that it appears their natural Element…."

Concluding an elaborate kava ceremony, a Hawaiian priest reverently offers a baked pig to Captain Cook, seated before holy images in a village near Kealakekua Bay. Unbeknown to the Englishmen, the Hawaiians saw in Cook the incarnation of Lono, their god of peace and bounty. Lono's symbol, a staff with cloth attached to it, resembled the masts and sails of Cook's ships; Cook had arrived during makahiki, *the season sacred to Lono; and tradition foretold the return of Lono by sea. At first the Hawaiians received Cook with awe and rites of adoration.*

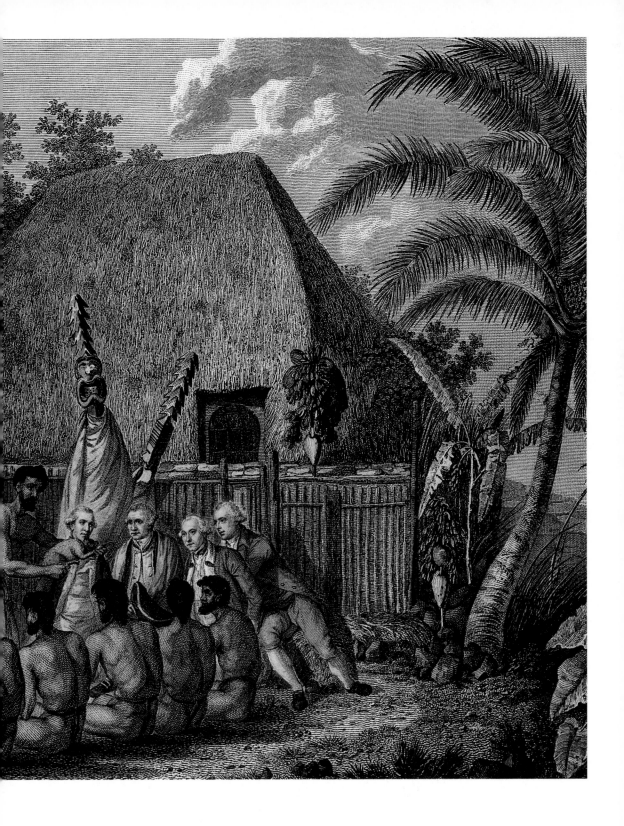

"These People pay the greatest attention to
Capt.ⁿ Cook....he was invested by them with the Title
and Dignity of Orono...a Character...
partaking something of divinity...."

and in strong gales on the night of the 7th the *Resolution*'s foremast was sprung—leaving her with only half her sails usable. For repairs Cook needed a sheltered harbor. Waimea Bay lay not only distant but also exposed to untrustworthy winds. He had struggled about for days before finding his harbor at Hawaii. In a choice of evils, prudence favored Kealakekua Bay, and with trepidation he turned south. "At 10," wrote King, "we bore away for Karakakooa, all hands much chagrin'd & damning the Foremast."

Early on the 11th the ships dropped anchor, and the carpenters quickly set to work. The Hawaiians were keeping their distance. Then Kalei'opu'u boarded Cook's ship and, according to Burney, "was very inquisitive . . . to know the reason of our return, and appeared much dissatisfied at it." This spirit of wary challenge seemed to infect most of the islanders; it darkened to

Carver Leonard Pipi fashions an image of Lono from ohia wood. Opposite, another ki'i, or idol, of Lono stands twice human size in an ancient pu'uhonua, a sacred place of refuge for breakers of tabu, or anyone else in danger of death.

menace and scorn. Thievery became epidemic. Clerke ordered forty lashes for a man who stole the armorer's tongs. A party getting water was threatened by "insolent" Hawaiians armed with stones. Cook ordered his men to load their guns not with the usual light shot, which would sting, but with combat ammunition: ball, which would kill.

Again the armorer's tongs were stolen from the *Discovery.* In the chase that resulted, Cook pursued the wrong man for miles in the wrong direction, derided by the crowd that misled him. Meanwhile, Edgar and Vancouver, trying to catch the same thief, got into a melee in which both were stoned. The reverence accorded the strangers had turned to contemptuous disdain.

That evening, wrote King, "the Captn expressd his sorrow, that the behaviour of the Indians would at last oblige him to use force; for that they must not he said imagine they have gained an advantage over us." If Cook said nothing of other moods, that was in character; he might well have been

PAGES 204-205: *Somber sunset engulfs ki'i and temple at Pu'uhonua-o-Honaunau, today a national historical park. After 18 days of ceremony and feasting at Kealakekua Bay, about 50 miles to the north, Cook set sail again. Heavy storms damaged his ship, however, and he limped back to the uncertain refuge of the bay.*

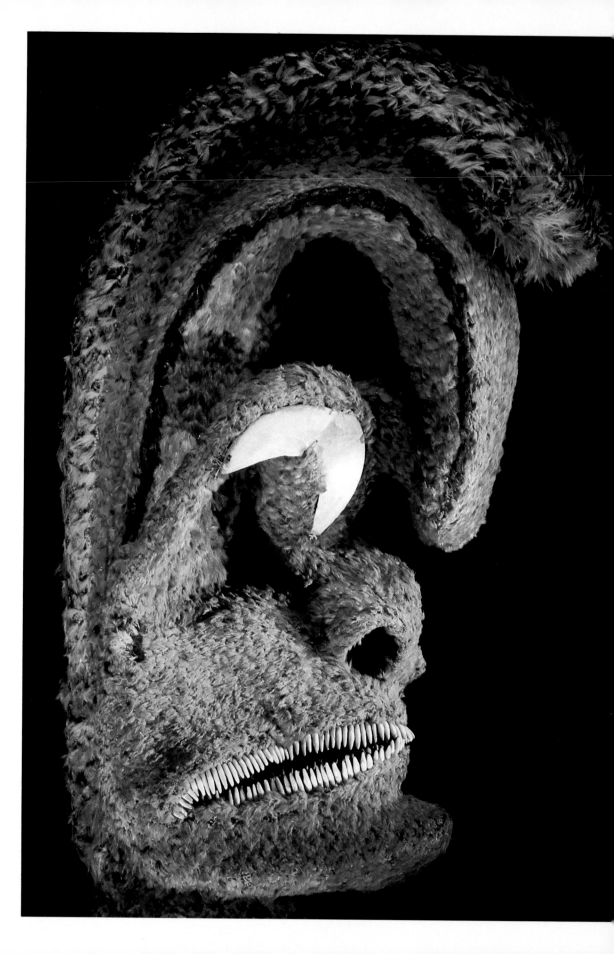

shocked and baffled by the Hawaiians' scorn, indignant at their treatment of his men, and deeply humiliated by his own wild-goose chase of the day. And recently he had shown signs of an unusually short temper.

Captain Clerke was also brooding over the turn of events: "our friendly connections having lull'd us into too great security."

Gentle breezes brushed the ships through a clear and pleasant night; and before dawn some Hawaiians stole the *Discovery*'s large cutter from under the noses of her night watch. Important for inshore exploration, this craft would be the chief lifeboat in emergency: Cook meant to retrieve it immediately. He knew a strategy that had worked on other islands—to take a chieftain hostage and hold him until the cutter was returned. Clerke, who had reported the theft, looked for hopeful signs: "we were as yet by no means upon bad terms" with the chiefs. Clerke had just left with his orders when James King found his commander, obviously disturbed, loading his double-barreled gun—one barrel with shot, the other with ball.

Cook had decided upon a show of force. He went ashore in his pinnace with Lt. Molesworth Phillips and nine of his marines, all armed. The captain had concluded, said Phillips, that gunfire would send the Hawaiians fleeing if trouble developed. One of Cook's small craft escorted him in and lay on call in the shallows, while other boats guarded the exits from the bay to keep canoes from leaving it. At a camp on the beach, King posted an armed guard and sought out the priest Koa, assuring him that Cook would let no one be hurt.

Staunch in his dignity, Cook marched into the village where Kalei'opu'u was staying, asked directions to his hut, and went straight to it. The marines clumped after him, virtually undrilled and certainly untested. The aged ali'i, just waking up, obviously knew nothing of the missing cutter and accepted Cook's invitation to visit the *Resolution*. He had two spirited young sons who had been spending hours on board with the captain and were eager to go along now; one scampered happily ahead to the pinnace.

As the rest started out, one of Kalei'opu'u's wives and two minor chiefs rushed up to detain him; the ali'i, said Phillips, "appear'd dejected and frighten'd" and simply sat down on the ground. A crowd of agitated Hawaiians, two or three thousand, had gathered; some had "war mats" or shields, some carried spears and daggers, some were picking up rocks. With Cook's approval Phillips ordered his marines to form a line at the water's edge, facing the crowd; it parted to let them pass. Assessing the crowd and the ali'i, Cook remarked to Phillips, "We can never think of compelling him to go on board without killing a number of these People."

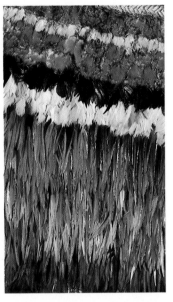

Grimace of dogs' teeth and a glinting eye of shell mark an image made for prayer to Lono, or Ku, god of war, or other deities. Feathers cover its three-foot frame of basketwork; they mingle on a ceremonial cape (above.) At first, Hawaiians gave Cook such treasures; but at his return, their mood darkened.

During this episode musket fire echoed from the south, where an officer on guard killed a chief who was trying to leave the bay. A companion of the victim's tried to report this calamity to Cook, with no success. Instead, news of it came rippling swiftly through the crowd. Cook was making his way slowly, deliberately, toward the waterline. Suddenly a man jumped in front of him, brandishing a stone and an iron trade-goods dagger either in deadly threat or insulting derision. Startled, Cook swung the muzzle of his gun toward this man and fired — fired the barrel loaded with shot. The thick war mat absorbed the pattern easily; the warrior himself was unharmed; the noise was as nothing.

*"another Indian ... drew out an iron Dagger
... & stuck it with all his force
into the back of his Neck ....
a fellow gave him a blow on the head
with a large Club and he was
seen alive no more."*

Impassioned multitude swarms to the attack on the morning of February 14, 1779. Cook, his gun empty, signals to his men as a Hawaiian brandishes a knife. Clubbed and stabbed, Cook stumbled into the surf and died under the blows of dozens of men. Tense and increasingly ominous clashes over several days led to this last, sudden, fatal confrontation.

"I could not observe the least fright it occasion'd"—the thought flashed through Phillips's mind as he foiled a dagger blow with the butt of his musket. Now the chief's son in the pinnace grew alarmed; the British seamen let him go; he swam to safety. Stones began to rain upon the shore party, a marine fell, Cook fired—and killed a man. Phillips fired, and the infuriated Hawaiians began a general attack, as overwhelming as a storm wave. Captain Cook's voice cut through the screams of the crowd, ordering the marines to fire and then shouting "Take to the boats!"

In "a most miserable scene of confusion" Phillips was knocked down and stabbed in the shoulder; he shot his attacker, scrambled to the water, and—though stunned by a stone—managed to swim to the pinnace. The men in the boats had opened a brisk fire that briefly checked the crowd and covered the escape of five marines; four lay dead on the rocks.

The *Resolution's* cannon were firing toward the distant turmoil.

Cook, his gun empty, signaled for the boats to come in. As he waved, one Hawaiian clubbed him from behind; another brought an iron dagger flashing down toward his neck. Stunned, the captain staggered into the surf and fell face down. The Hawaiians descended on him and held him under and with repeated blows killed the man they worshiped as a god.

Fluttering light from a net fisherman's torch spills across a bronze plaque dedicated to the memory of Captain James Cook. His remains lie in the embrace of the waters of Kealakekua Bay, but his legacy endures throughout his beloved Pacific Ocean.

Epilogue:
"A Man of Unparalleled Success"

News of Cook's death brought grief home to many. Walker, Palliser, Sandwich, Banks, Solander, Stephens — all were desolated. Elizabeth Cook and her sons were heartbroken. George III, it was said, shed tears. Even crusty Johann Reinhold Forster was moved to eulogy: "If we consider his extreme abilities, both natural and acquired, the firmness and constancy of his mind, his truly paternal care for the crew entrusted to him, the amiable manner with which he knew how to gain the friendship of all the savage and uncultivated nations, and even his conduct towards his friends and acquaintance, we must acknowledge him to have been one of the greatest men of his age, and that Reason justifies the tear which Friendship pays to his memory."

The loss of Captain Cook stung the hearts of people in lands beyond his own. In his competent, practical way, he had come to symbolize the humane explorer. He had taken mankind to the limits of the globe, and had stirred imaginations everywhere.

On no one did the weight of his passing fall more heavily than on Charles Clerke. Debilitated by disease, bereft of a friend, Clerke was thrust into command in daunting circumstances. But he had learned well from Cook's example. Against demands for blood, Clerke imposed restraint. His men hungered to pillage and destroy; Clerke set them to work on the ships. The Hawaiians wavered between abject fear and haughty aggression; Clerke treated them with firm dispassion. A man of less merit might have resolved on a course of revenge which not only would have deepened the tragedy but also might well have doomed the expedition. Clerke masterfully manipulated the passions of his men and the Hawaiians, and chaos was circumvented.

Clerke's one demand was the return of his captain's body. Island tradition precluded this. When a great man died, his body was burned and the bones distributed as relics to rulers around the island. Cook was accorded this tribute. Clerke learned of it during the night of the 15th, when Keli'ikea came secretly, at the risk of his own life, to bring a token of friendship: flesh rescued from the pyre.

Finally, almost a week after the killing, a sombre procession wound from the heights toward Kealakekua Bay, bringing offerings of food and a small bundle wrapped in tapa cloth. In the *Resolution*'s great cabin, Clerke unwrapped the bundle and regarded the remains of his revered friend: scalp, skull, bones of arms and legs, hands. The hands, preserved with salt, proved that these relics indeed were Cook's — the old scar from the powder-horn explosion in Newfoundland traced a jagged line along the right hand. Surgeon's mate David Samwell, present in the great cabin, spoke for those "who looked upon Captn Cook as their father & whose great Qualities they venerated almost to adoration ... what their feelings were upon the Occasion is not to be described."

In the evening of the next day, Clerke consigned the remains of James Cook to the waters of Kealakekua Bay, "with all the attention and honour we could possibly pay it in this part of the World." The following day, with light breezes to convey them, the *Resolution* and *Discovery* left the island of Hawaii. "An universal gloom, & strong Sentiments of Grief & Melancholy were very observable throughout all ranks ... on our Quitting this Bay without our great & revered Commander," wrote Trevenen.

Professionally, but without spirit, Clerke guided his men among the Hawaiian Islands, then north to the Arctic to search again for the Northwest Passage. Conditions were harsh, and in spite of devoted efforts the expedition was forced to abandon its mission. At sea near the Russian port of Petropavlovsk on the Kamchatka Peninsula — the first stop on the long voyage home — the benevolent and personable Charles Clerke died of consumption. Lts. John Gore and James King finally brought the ships to anchor in home waters on August 22, 1780 — four weary years after departure.

The waves of excitement and praise stirred by the former homecomings were becalmed now. London had learned of Cook's death already, by a letter forwarded from Petropavlovsk. Moreover, Great Britain was preoccupied with the War of American Independence, now in its fourth year.

Still, the esteem in which Cook was held was made abundantly plain. The Royal Society struck a medal in his honor. King George III granted a posthumous coat of arms: a map of the Pacific showing the voyages; the polar stars; a wreath of laurel and palm; mottoes in Latin proclaiming *Circa orbem* — "Around the globe" and *Nil intentatum reliquit* — "He left nothing untried."

Sir Hugh Palliser — who suppressed Cook's criticism of the Navy Board—erected an obelisk dedicated "To the memory of Captain James Cook, The ablest and most renowned Navigator this or any country hath produced. . . . Traveller! contemplate, admire, revere, and emulate this great master in his profession; whose skill and labours have enlarged natural philosophy; have extended nautical science; and have disclosed the long concealed and admirable arrangements of the Almighty in the formation of this globe. . . ."

Most appropriately, the government gave Elizabeth Cook a handsome pension. Of her sixteen years of marriage to James Cook, she had been with him less than five. She had lost three of their six children. And now she had lost her husband in a tragedy she could barely comprehend. In 1780 young Nathaniel Cook died in a shipwreck. In 1793 Hugh succumbed to a fever. And the next year James, the eldest son, was killed under mysterious circumstances off the Isle of Wight. A stalwart woman, Mrs. Cook lived to the age of 93; she passed away peaceably in 1835 — a full 56 years after her husband had been killed. To the end she revered him as a paragon to whom others could compare but poorly.

In many ways her judgment was apt. If not a man of boundless vision, James Cook was a man of unparalleled success. He mapped tens of millions of square miles; he discovered intriguing new peoples; he vastly expanded the horizons of the British Empire.

I asked Michael Hoare for a historian's summary of James Cook and his accomplishments. With a sigh, Dr. Hoare settled back in his chair. "That's almost an impossibility," he said. "The legacy of Captain Cook is so vast and so rich that it's difficult to tie it together in one thought. However, I will say that the only thing comparable in impact on science and in importance to the entire world since Cook's voyages was man's landing on the moon. The successes and results of these two incredible series of missions, separated by 200 years, transformed the way man viewed himself and his world."

One tranquil February evening, I sat on a hummock of black, crumbly lava at the edge of Kealakekua Bay—within a few feet of the spot where Cook was killed. A fireball sun hovered near the horizon, dappling the lavender waters with droplets of blood red. The tide murmured at my feet, and a warm breeze, musky with the scent of the sea, tugged gently at my hair.

For nearly a year I had pursued the wraith of Captain Cook around the world. That path had led me finally, conclusively, to Kealakekua Bay. I had come to know him well, I thought, but I still yearned to know him better.

I wished that, for just a few hours, I could have been Joseph Banks, or Charles Clerke, or James King. I would have walked the decks with Cook at midnight, listening to the creak of the rigging, the soft chants of the sailors, the gentle wash of the sea against the wooden hull. Or I would have sat with him in the great cabin discussing navigation or helping him record the intricacies of a ceremony we had witnessed. Or I would have landed with him on a strange beach and watched him befriend a group of wary islanders with his lack of bluster, his disciplined patience, his gentle ways.

James Cook, I mused, was a man who changed the face of the world. And perhaps his death in Hawaii was fitting; he had mastered a third of the globe, and what was left for him? It would have been unseemly, shameful, for him to die of old age in some dim bedchamber in London.

As I pondered these themes, the tide began to turn in Kealakekua Bay. Slowly, wave by wave, the sea lapped higher against a bronze plaque commemorating Cook's death, and soon the words were blurred. The sun faded below the horizon and left me in purple gloaming. Hardly breathing in the brittle stillness, I rose and gazed out across the bay. Somewhere in those waters, I reflected, the bones of Captain James Cook still rest, appropriately, in the embrace of his Pacific Ocean.

Acknowledgments and Illustrations Credits

The Special Publications Division is grateful to the individuals and organizations named or quoted in the text, and to those listed here, for their generous cooperation and assistance during the preparation of this book: Douglas Arnold, William S. Ayres, Nick Beck, Zorro A. Bradley, Colin Chalmers, Francis Fay, Robin Fisher, J. Geoffrey Graham, Robert J. Gregory, Dora Harrington, Jack Harter, Lisle Irwin, Tua John, Carol Jopling, Ota Joseph, Adrienne L. Kaeppler, Jay Luvaas, Clarence Maderos, Russell Marriott, Patrick C. McCoy, J. Claude Michaud, Jennifer Moseley, Anthony Murray-Oliver, Bob Nelson, Lowana Nierpiken, George Niitani, Bob and Kath Paul, Michel Pichon, Guy Rauzy, Bruce Reed, Saul H. Riesenberg, Richard Rosenblatt, Tiare Sanford, Jerry Y. Shimoda, David Simmons, Dan Strickland, Robert C. Suggs, 'Apai Tonga, Sir James Watt, Lionel J. Willis; the Embassies of Australia, France, and New Zealand; the National Maritime Museum, Greenwich, England, and the Smithsonian Institution.

In particular, the division extends its thanks to those listed here for their courtesy in making artifacts and works of art available for illustration. The hei-tiki on page 73 is reproduced by gracious permission of Her Majesty the Queen. The drawings of fishes by Alexander Buchan and Sydney Parkinson (p. 57), of plants by Parkinson (p. 58, p. 59), and a sketch by George Forster (p. 107), appear by permission of the Trustees of the British Museum (Natural History); the rendering of *Eugenia suborbicularis* (p. 58) is by F. P. Nodder and that of *Blandfordia nobilis* (p. 90) by James Miller, both after Parkinson. Cook's chart (p. 27) is reproduced by permission of the Controller of H. M. Stationery Office and of the Hydrographer of the Navy; and the painting (p. 35) by Buchan, Add. MS. 23920 F.14, by permission of the British Library, London.

From the Dixson Library, Sydney, Australia: pp. 96-7, engraving, hand colored, by Will Byrne after lost original attributed to Parkinson or to H. D. Spöring, in John Hawkesworth, *An Account of the Voyages . . .*, London, 1773. P. 143, watercolor by John Webber, "A View of Christmas Harbour in Kerguelen's Land." P. 182, watercolor by Webber, "Sea Horses." P. 194, watercolor by Webber, "Tereeboo [Kalei'opu'u], King of Owhyee, bringing presents to Capt. Cook." Pp. 200-201, engraving, hand colored, by Samuel Middiman and John Hall after Webber, "An Offering Before Capt. Cook, in the Sandwich Islands." From the Mitchell Library, Sydney: P. 8, watercolor by Henry Roberts. P. 105, watercolor by William Hodges, "The *Resolution* and *Adventure* . . . Taking in Ice for Water. Lat. 61°S." Jan. 1773. P. 208, engraving, hand colored, by F. Bartolozzi and W. Byrne after Webber, "The Death of Captain Cook," photographed by James A. Sugar.

From the National Maritime Museum: pp. 6-7, oil by Nathaniel Dance, "Capt. James Cook, R.N., F.R.S." P. 108-109, oil by Hodges, "View of Cape Stephens." P. 113, oil by Hodges, "The Monuments of Easter Island."

From the National Geographic Society Library, Rare Books Collection: p. 94, detail, engraving by J. Cheevers, in Hawkesworth, *An Account of the Voyages. . . .* P. 97, engraving by T. Chambers, in Sydney Parkinson, *A Journal of a Voyage to the South Seas*, London, 1773. P. 160, engraving by W. Sharp after Webber, "A Night Dance by Women in Hapaee," in James Cook, *A Voyage to the Pacific Ocean*, London, 1784.

From the Peabody Museum of Archaeology and Ethnology, Harvard University: p. 172, watercolor by Webber, "A Man of Nootka Sound." P. 173, Webber, "The Inside of a House in Nootka Sound." Both, April 1778; photographed by Hillel Burger.

From the Hocken Library, Dunedin, New Zealand: p. 56, detail, mezzotint engraving by J. R. Smith after oil by Benjamin West. The Bernice P. Bishop Museum, Honolulu: p. 60, mourning dress. The University Museum of Archaeology and Anthropology, Cambridge, England: p. 61, carving. The Alexander Turnbull Library, Wellington, New Zealand: p. 72, engraving by T. Chambers, in Parkinson, *Journal*, 1784 printing. Private Collection of Peter Rheinberger: p. 106, oil by J. F. Rigaud. Government House, New Zealand: p. 161, oil by Nathaniel Dance, "Captain Charles Clerke, R.N." Museum of Mankind, London: p. 206: feather image. Museum für Völkerkunde, Vienna, Austria: p. 207, feather cape.

Additional Reading

The reader may wish to consult the National Geographic Society Index for related articles, and to refer to the following books.

Primary Sources: J. C. Beaglehole, ed., *The Journals of Captain James Cook* and *The Endeavour Journal of Joseph Banks 1768-1771*; James Burney, *With Captain James Cook in the Antarctic and Pacific*, ed. Beverley Hooper; George Forster, *A Voyage Round the World in His Britannic Majesty's Sloop, Resolution*, London, 1777; John Ledyard, *John Ledyard's Journal of Captain Cook's Last Voyage*, ed. James Kenneth Munford; Anders Sparrman, *A Voyage Round the World*, tr. Huldine Beamish and Averil Mackenzie-Grieve.

Secondary Sources: Terence Barrow, *Captain Cook in Hawaii*; J. C. Beaglehole, *The Exploration of the Pacific* and *The Life of Captain James Cook*; Hugh Cobbe, ed., *Cook's Voyages and Peoples of the Pacific*; Daniel Conner and Lorraine Miller, *Master Mariner: Capt. James Cook and the Peoples of the Pacific*; Ernest S. Dodge, *Northwest by Sea*; Robin Fisher and Hugh Johnston, eds., *Captain James Cook and His Times*; Howard T. Fry, *Alexander Dalrymple and the Expansion of British Trade*; Michael E. Hoare, *The Tactless Philosopher: Johann Reinhold Forster*; Jesse D. Jennings, ed., *The Prehistory of Polynesia*; Adrienne L. Kaeppler, "Artificial Curiosities"; Gavin Kennedy, *The Death of Captain Cook*; T. C. Mitchell, ed., *Captain Cook and the South Pacific: The British Museum Yearbook 3*; Alan Moorehead, *The Fatal Impact*; Anthony Murray-Oliver, *Captain Cook's Artists in the Pacific* and *Captain Cook's Hawaii*; Colin Newbury, *Tahiti Nui*; Douglas L. Oliver, *Ancient Tahitian Society*; Andrew Sharp, *The Discovery of Australia*; R. A. Skelton, *Captain James Cook After Two Hundred Years*; Bernard W. Smith, *European Vision and the South Pacific, 1768-1850*; Tom and Cordelia Stamp, *James Cook: Maritime Scientist*; Thomas Vaughn and Anthony Murray-Oliver, *Captain Cook, R. N.: The Resolute Mariner*; Alan Villiers, *Captain James Cook*.

Index

Boldface indicates illustrations; *italic* refers to picture legends (captions)

Aboriginals 83, 90, **97**, **99-101**
Admiralty, Board of 25, 27, *31*, 34, 35, 36, 68, 83, 99, 104, 113, 142, 211; orders 36, 41, 69, 143, 169, 185
Adventure 104, **105**, 107, 113, 116, 118, 143, 154; separations from *Resolution* 112, 118
Adventure Bay, Tasmania 153
Ahu (sacred area) *112*, **113**, *115*, 119
Aireyholme Farm, near Great Ayton, England 12, **17**
Aitutaki (island), Cook Islands **140-141**, **144**, **145**, **152-153**; coral lagoon **148-151**
Alaska 9, 169, 172-173, **174-175**, **176**, 177-178, **178**, **179**
Aleuts 178, *178*, **179**, 184-185
Améré Island, New Caledonia **138**
Anaterea, Mareta **1**, **144**, **145**, 153
Anderson, Susan J. 169
Anderson, William 143, 144, *148*, 154, 161, 172-173, 185
Anga 'Unga family *159*, 160

A Note on Quotations

In chapter titles and elsewhere, quotations rendered in calligraphy follow the author's text, the wording of his informants (Greg Doyle, **p. 11**, **p. 30**; Leon Tepa, **p. 53**), or that of Cook and men of his expeditions. For these, spelling and style follow Beaglehole's edition of the *Journals*, or his *Life*. **P. 6:** Cook, Jan. 30, 1774.

Chapter 1, p. 27: Admiralty order, in *Life*, p. 67.

Chapter 2, p. 35: Cook, Jan. 16, 1769. **P. 38:** Robert Molyneux, April 22, 1769. **P. 46:** Cook, April 13, 1769.

Chapter 3, p. 73: John Gore, Oct. 9, 1769. **P. 74:** Cook, Nov. 13, 1769. **P. 80:** Sydney Parkinson, In *Life*, p. 219. **P. 88:** Cook, May 6, 1770. **P. 92:** Cook, June 10, 1770. **P. 94:** Cook, June 11, 1770.

Chapter 4, p. 108: Charles Clerke, May 18, 1773. **P. 113:** Cook, March 17, 1774. **P. 114:** Clerke, March 1774. **P. 123:** Clerke, April 1774. **P. 124:** William Wales, Aug. 8, 1774. **P. 135:** Cook, Aug. 10, 1774. **P. 137:** Wales, Aug. 13, 1774. **P. 138:** Wales, Sept. 23, 1774.

Chapter 5, p. 144: Gore, April 1777. **P. 149:** William Anderson, April 14, 1777. **P. 150:** James King, July 1777. **P. 159:** Cook, Oct. 26, 1777. **P. 162:** Cook, Jan. 23, 1778.

Chapter 6, p. 167: Clerke, Feb. 11, 1778. **P. 170:** Cook, March 7, 1778. **P. 177:** Clerke, June 26, 1778. **P. 182:** Cook, Aug. 19, 1778.

Chapter 7, p. 196: King, Dec. 2, 1778. **P. 199:** Clerke, Jan. 1778. **P. 201:** David Samwell, Jan. 19, 1779. **P. 208:** Samwell, Feb. 14, 1779.

Antarctic 9, **105,** 110, 112, 118; map **102;** *see also* Kerguelen Islands

Aorangi (mountain), New Zealand *see* Cook, Mount

Araucaria columnaris ("Cook pine") 6, 8, **138**

Arctic 169, 177, *177,* 178, **180-183, 186-189,** 210; map **166**

Astronomy 27, *31, 34,* 36, *36,* 44, 56, *70,* 73; observatories **46,** 53, 113; transit of Venus 34, *36,* 53, 57

Atiu (island), Cook Islands 154

Atlantic Ocean 44, 99, 107, 110, 138, 144; pre-Cook exploration 107, 110, 143; *see also* Canada; North Sea

Attacks by natives 72, 73, 79, 118, 129-130, 131, 202, 207-208, **208;** Tongan plot 155, *160*

Austral Islands, French Polynesia *40,* **62-63,** 68; *see also* Rurutu *and* Tubuai

Australia 9, **16,** 69, 82-83, **84-87,** 87, **88-89, 92-93;** maps **66, 83, 94**

Balcom, Sandy 22

Banks, Joseph 36, 45, 56, **56,** 57, 68, 69, 73, 79, 82, 87, 98, 104, 105, 107, 143, 144, 210, 211; collection 83, **91;** journals 72

Banksia 82, **88**

Bark cloth 57, 117, **156-157,** 162, 194, 210; mourning dress **60**

Batavia, Java 98-99

Bayly, William 107, 117, 161, 168

Beaglehole, John Cawte 9, 105, 131, 155

Bering, Vitus 172, 184

Bering Sea 172, 178, **182-183**

Bering Strait 9, 173, **180-183**

Bevis, James 27

Birds 79, **106,** 107, 113, **180;** lorikeet **87;** penguins **107, 143,** 153; seagulls **10-11, 79**

Bligh, William 143, 178, 193, *195*

Blowholes **108,** 117

Boats *see* Canoes; Umiaks

Bonne Bay, Newfoundland **28-29**

Bora Bora, Society Islands **54-55,** 68

Boswell, James 142

Botany 6, 8, 56, 69, 72, 73, 82-83, *88, 90,* 113, *148;* drawings **58-59, 90, 91**

Botany Bay, Australia 82-83, *88, 90*

Bougainville, Chevalier de 46, 61, 130

Bouvet, Lozier 107, 110

Breadfruit **40, 159**

Brett, Cape, New Zealand **76-77,** *79*

Brigus South, Newfoundland 26

Buchan, Alexander 36, *56,* 57; paintings by **35, 57**

Burney, James 117, 118, 143, 169, 193, 202

Canada: Maritime Provinces 20, 22-23, **22, 24, 25,** 25-27, *27,* **28-31;** West Coast 168-169, **172, 173,** 178

Canary Islands, Atlantic Ocean 144

Cannibal Cove, New Zealand 79

Cannibalism 79, 118, 120

Canoe races **50-51,** 61, 64, **198-199**

Canoes: 53, 61, *61,* 64; double **50-51,** 129, **194;** outrigger **36, 42, 62-63, 126-127, 150-151, 190-191, 198-199;** sailing **62-63, 150-151,** 162, **194**

Cape Town, South Africa 99, 107, 144, 153

Cat-built barks **8,** 13, 36, 104, 143

Charting 211; Canada 23, 25-27, *27, 31;* South Pacific 57, 61, 64, 69, 73, 83, 112-113, 119, 132; North Pacific 168, *169*

Charts and maps: North America 23, 26, 27, **27,** *31,* 36; North Pacific 178; South Pacific 69, **94,** 99, 104

Chirikov, Alexei 172, 184

Christmas celebrations 44, 79, 105, 118, 138, *143,* 161

Chronometers 112, **116,** 129

Chukchi Sea, Arctic 177, **180-181, 188-189**

Circumcision ceremony: Tanna (island), New Hebrides 132, **133-137**

Clerke, Charles 99, **105,** 107, 110, 113, 116, 118, 120, 129, 144, *198,* 210, 211; command of *Discovery* 143, 153, 160, 161, **161,** 168, 172, 195, 202, 207; consumption 144, 161, *161,* 192, 210

Colnett, Cape, New Caledonia 138

Colville, Alexander, Lord 23, 25, 36

Cook, Elizabeth Batts (wife) 25, 26, 27, 36, 104, 142, 210, 211

Cook, Elizabeth (daughter) 27, 104

Cook, Hugh (son) 144, 211

Cook, James: apprenticeship in colliers 13, 20; character and temperament 9, 12, 142, 155, 169, 185, 192-193, 210; death and disposition of remains 208, 210; diplomacy 9, 53, 56, 61, 64, 113, 129, 131, 132, 154, 169, 211 *see also* Trading; geographical namesakes 9, 79, 82, 90, 117, *138,* 172; hand injury 26, 210; honors 104, 142, 144, 195, 211; illness 119, 138, 155; journals 9, 69, 72, 118, 142, 143, 144, 193; letters 104, 144, 184; navigational expertise 6, 8, 9, 20, *171;* portrait **6-7,** 143; presentation to His Majesty 104, 142; punishment of islanders 72, 73, 129, 155, 202, 207-208, hostage taking 56, 64, 207; rules for dealing with islanders 53, 129, 162, 192; scientific investigation *see* Astronomy; Botany; Zoology; ship's discipline 45, 46, 72, 98, 116, 161; surveying 23, 25-27, *27, 31, see also* Charting; Charts and maps; youth 12-13, 17, baptismal record **17**

Cook, James (son) 26, 104, 142, 211

Cook, Nathaniel (son) 27, 104, 142, 211

Cook, Mount, New Zealand **80-81,** 82

Cook Inlet, Alaska 172, **174-175**

Cook Islands, South Pacific 117, **140-141,** 144, **144-153,** 154

"Cook pines" 6, **8, 138**

Cook Strait, New Zealand *69,* 79, 113, 116, 118, 138

Cooktown, Australia 90

Coral 6, 9, 116, 117, 138, *146,* **148-149,** 154, 160; blowholes **108;** Great Barrier Reef 83, 87, 90, **92-95,** 98

Dalrymple, Alexander 35, 45, 118, 138

Dance *38,* 57, *72,* **110-111,** 116, **140-141, 153,** 155, 160, **160,** 194

Diomede Islands, Alaska-U.S.S.R. 173, 177, **180-184,** *185*

Discovery 143, **143,** 144, 155, 162, 168, 177, 192, 194, 195, 207, 210; repairs 144, 160; separation from *Resolution* 193

Disease: consumption 99, 144, 161, 210; malaria and dysentery *58,* 98-99; venereal *39,* 61, 162, 192; *see also* Scurvy

Dolphin, H.M.S. 34, 53

Donovan, Ken 22

Doyle, Greg 26

Duc d'Aquitaine (merchantman) 20

Dusky Sound, New Zealand **66-67,** 82, 112

Eagle, H.M.S. 20

Easter Island **102-103, 112, 113-115,** 119-120

Eclipses, solar 27, 110

Edgar, Thomas 153, 202

Elliott, John 113, 118, 130

Endeavour 36, 44, 46, 53, 68, 72, 73, 82, 83, **97,** 98, 99, 142; repairs 57, 64, 87, *97,* 99, 104; stranding 87

Endeavour Reef, Australia 87, **94-95**

Endeavour River, Australia 90, **96-97,** *99,* **100-101**

England 9, **10-11,** 12-13, **14-15,** 17, **17-19,** 20, **21,** 25, 26; homecomings 104-105, 107, 142-144, 210-211

English Channel 20

Equator, crossing of 44, 161

Erromango (island), New Hebrides 131

Eskimos 173, 177, **180-187,** 185

Eucalypts 82

Europeans' impact on peoples of the Pacific *39, 49,* 57, 61, 132, 153, 162, 195

Everard, Cape, Australia 82

Fairweather, Cape, and Mount, Alaska 169

Fatu Hiva (island), Marquesas **122-123,** 129

Figueroa, Gonzalo 119

Fiji Islands 69, 154, 161

Finau (Tongan chief) 155

First voyage (South Pacific, 1768-1771) 34-104; crew 36, deaths *58,* 99; purpose 34, *34,* instructions 36, 41

Firth, Ronald M. 12

Fishes *57,* 113, **149,** 154

Fishing **146-147;** cod 22, 26, **30-31;** with spear and arrow **126-127**

Flattery, Cape, Washington 168

Forster, George **106,** 107, 113, 119, 131, 138

Forster, Johann Reinhold **106,** 107, 113, 138, 210

Foulweather, Cape, Oregon **166-167,** 168, **170-171**

Franklin, Benjamin 142, 172, 178

Freelove (collier) 13, 36

French Polynesia 56, 68; *see also* Austral Islands; Marquesas Islands; Society Islands; Tuamotu Archipelago

Freshwater Cove, Newfoundland 26, **30-31**

The Friendly Islands *see* Tonga

Fry, Howard T. 34, 35

Furneaux, Tobias 105, 107, 112, 113, 116, 117, 118, 143, 153

Gallop, Graham 90

George III, King (England) 72, 104, 142, 143, 210, 211

Gibson, Samuel 64

Good Hope, Cape of 99, 143

Good Success, Bay of, South America 44

Gore, John 73, 87, 143, 144, 153, 172, 210
Grand Banks, Newfoundland 31
Great Barrier Reef, Australia 83, 83, 87, 90, 92-95, 97, 98
Green, Charles 36, 57, 70, 73, 98, 99
Greenwich Hospital 142
Grenville (schooner) 26, 27
Gromoff, Ishmael 179, 184-185

Haleakala Crater, Maui, Hawaii 192
Halley, Edmund 34
Harrison, John: chronometer 112
Haslam, D. W. 9, 26, 69
Hawaii (island), Hawaii 192-193, 196-197, 200-201, 202, 204-205, 207-208, 209
Hawaiian Islands, North Pacific 9, 161-162, 162, 163-165, 190-209, 192-210; map 193
Hawaiians 161, 190-191, 193, 194, 198-202, 202, 208, 209; chiefs 193-195, 194, 207, 210; deities 194, 198, 207 see also Lono; feather capes 195, 207
Hawkesworth, John 142
Hawkins, Clarence, and brothers 30
Hei-tiki (ornament) 72, 73
Heiau (sacred area) 162, 193-194
Hervey Island, Cook Islands 117, 154
Hicks, Zachary 87, 98, 99
Hiva Oa (island), Marquesas 120, 120, 129
Hoare, Michael E. 8, 142, 211
Hodges, William 107; paintings by 108-109, 113
Hogs, wild ("Captain Cookers") 154
Holland, Samuel 23
Homecomings 99, 104, 142, 210-211
Hope, Point, Alaska 187
Horn, Cape, South America 44-45, 138
Hot springs: Tanna (island), New Hebrides 128
Huahine (island), Society Islands 68, 143, 161

Ice 45, 105, 107, 110, 112, 118, 177, 178, 180-181, 186-187, 188-189; drinking water from 105, 110
Icy Cape, Alaska 177, 187
Ignaluk, Little Diomede Island, Alaska 177
Indian Ocean 110, 143; map 140
Indians: name for natives 53; North America 168-169, 172, 173; South America 35, 44
Ismailov, Gerassim Gregoriev 178, 184
Ivory, walrus-tusk 177, 184
Iyahuk, Albert 177
Iyapana, John 173, 177

Jakarta, Java, Indonesia see former name, Batavia

Kalekale, Tevita 158
Kamchatka Peninsula, U.S.S.R. 184, 210
Kangaroos 83, 84-86, 87, 90
Kauai (island), Hawaii 162, 162-165
Kava 117, 132, 134, 160, 195, 200
Kealakekua Bay, Hawaii 193-195, 194, 195, 200, 202, 202, 207-208, 209, 210, 211
Kendall, Larcum: chronometer 112, 116, 129

Kerguelen Islands, Indian Ocean 110, 143, 153
King, James 143, 151, 155, 168, 169, 173, 178, 195, 202, 207, 210, 211
Kippis, Andrew 142

Labrador 26
Latitude, calculation of 110, 113, 129
LeBlanc, Denise 24
Ledyard, John 169
Lifuka (island), Ha'apai Group, Tonga 155, 158-159, 160, 160
Lind, James 45
Little Diomede Island, Alaska 173, 177, 180-184
Livestock: ship's provisions 44; transported to Tahiti as gift 143, 144, 153, 154, 160
Lizard Island, Australia 90
London, England 13, 20, 26, 99, 104, 142, 210-211
Longitude, calculation of 27, 44, 73, 110, 112, 113, 116, 129
Lono (Hawaiian deity) 193-194, 200, 202, 203, 207
Louisbourg, Cape Breton Island, Canada 20, 22, 22, 23, 24, 25

Madeira Islands, Atlantic Ocean 44, 107
Maeva, Teeiau 1, 144, 145
Malekula (island), New Hebrides 130-131
Mangaia (island), Cook Islands 146-147, 151, 154
Maori 71, 72, 72, 73, 73, 79, 79, 81, 82, 109, 112, 113, 154, 161
Maps: Antarctic 102; Arctic 166; Australia 66, 83, 94; Hawaii 193; Indian Ocean 140; New Zealand 66, 69; North Atlantic 10; North Pacific 166, 190; South Pacific 102; Tahiti 45; Tonga 155; World 32; see also Charts
Marae (sacred area) 64
Marquesans 121
Marquesas Islands, French Polynesia 120, 122-123, 129, 161, 162
Marton-in-Cleveland, England 12, 17, 132
Matavai Bay, Tahiti 37, 42-43, 46, 53, 64, 116, 129, 161
Maui (island), Hawaii 192, 193
Mauna Kea (volcano), Hawaii 192, 196-197
Mauna Loa (volcano), Hawaii 192
Melanesia see New Caledonia; New Hebrides
Melanesians 130-132, 138
Melbourne, Australia: Fitzroy Gardens, Cook's parents' home transported to 16
Mercury Bay, New Zealand 70-71, 73-74
Metal: importance to natives 53, 101, 173, 192; see also Nails
Milligrock, Dwight 177, 184
Missionaries 49, 185
Moai (statues) 102-103, 112, 113-115, 119
Mogg, Sam 177
Molyneux, Robert 53, 57, 99
Moorea, French Polynesia 36, 40, 42-43, 155, 161

Moortgat, Paul-R. 56
Motuara Island, New Zealand 78
Motuarohia Island, New Zealand 78
Mount Cook National Park, New Zealand 80-81, 82

Nails 38-39, 53, 74, 117, 129, 130, 192
Na Pali coast, Kauai (island), Hawaii 162, 163
Navigation 6, 8, 9, 20, 112, 171; instruments 112, 116, 129; see also Latitude; Longitude
New Caledonia (island), South Pacific 2-3, 6, 138, 138, 139
New Hebrides (islands), South Pacific 124-128, 130-132, 133-137
New Zealand 9, 69, 66-67, 70-71, 72, 74, 76-77, 108-109, 112, 113, 153-154; circumnavigation 82; map 66, 69; Southern Alps 80-81, 82
Newfoundland, Canada 25, 26, 27, 27-31
Niihau (island), Hawaii 162
Niue (island), South Pacific 130
Nomuka (island), Tonga 130, 155
Nootka Indians 168-169, 172, 173, 178
Nootka Sound, Vancouver Island, Canada 168-169
Norfolk Island, South Pacific 138
North America: east coast 20, 22-23, 22, 24, 25, 25-27, 27, 28-31; west coast 166-167, 168-169, 170-171, 172-173, 174-176, 177-178, 179-183, 184-185, 186-189
North Sea 9, 12, 12, 13
Northernmost landfall 177, 187
Northumberland, H.M.S. 23, 25
Northwest Passage: search for 143, 154, 168, 169, 172, 185, 210
Nova Scotia, Canada 23, 25

Omai (Huahinean) 143, 153, 154, 161
Oregon coast 166-167, 168, 170-171
Otaheite see Tahiti
Oysters 74, 74, 75

Pacific Ocean: pre-Cook exploration 9, 34, 41, 46, 53, 56, 61, 69, 72, 74, 82, 119, 120, 130; North Pacific 168, 172
Palliser, Hugh 20, 26, 36, 142, 193, 210, 211
Palmerston Atoll, South Pacific 148, 154
Pandanus 111; weaving 41, 68
Paowang (Tannese) 131-132
Papeete, Tahiti 56
Parkinson, Sydney 36, 46, 56, 61, 73, 99; paintings by 57-59, 72, 83, 89, 90, 91, 97, 97, 161
Patten, James 119
Pembroke, H.M.S. 20, 22-23
Phalp, Mark 17
Phillips, Molesworth 207-208
Pickersgill, Richard 72, 98, 99, 107
Pickersgill Harbour, New Zealand 112-113
Pigs 134, 159, 160, 193, 200
Pipi, Leonard 202
Plymouth, England 36, 107, 144
Point Venus, Tahiti 36, 53, 56, 64
Polynesians 9, 61, 115, 119, 120, 161-162; bark-cloth mourning dress 60; migrations 161, 162
Poto, Vaine Rere Tangata 154
Poverty Bay, New Zealand 69, 73

Prince of Wales, Cape, Alaska 173, 177
Promiscuity *38-39,* 53, 57, 117, 130, 192, 195
Providential Channel, Australia 98
Provisions 35, 44, 45, 72, 107, 129, 132, 153, 154, 155, 177, *183,* 185, 192
Prowd, Mary 13, 105
Pu'uhonua-o-Honaunau National Historical Park, Hawaii *202,* **204-205**

Quebec City, Quebec, Canada 20, 22, 22, 23
Queen Charlotte Sound, New Zealand 79, 112, 113, 118

Rapu, Sergio 119-120
Ravenhill, John 98, 99
Resolution 6, 8, **8,** 104, 105, **105, 108, 143,** 192, 207, 210; in hazard 116; repairs 138, 142, 144, 160, 169, 178, 202; separations from *Adventure* 112, 118; separation from *Discovery* 193
Resolution, Port, Tanna (island), New Hebrides **128,** 129, 132
Riou, Edward 192
Roberts, Henry: painting by **8**
Roggeveen, Jacob 119
Royal Navy 9, 20, *22,* 23, 25, 27, 36; conditions of service 20; corruption 144, 193; *see also* Admiralty
Royal Society 27, 31, 34, 35, 36, 41, 142, 211; Copley Medal 144, 172
Rurutu, Austral Islands **41, 52,** 68, 69
Russians in Alaska 178, *178,* 184-185, 210; Orthodox Church **178, 179**

St. Lawrence River, Canada 20, 22, *22,* 23, 25
St. Pierre and Miquelon (islands) 25-26
Salmon, Tutaha 61, 64
Samoa Islands, Central Pacific 154, 161
Samwell, David 169, 185, 195, 210
Sanderson, William 12, 13
Sandwich, Earl of (John Montagu) 104, 105, 142, 144, 193, 210
Sandwich Islands *see* Hawaiian Islands
Scurvy 45, 116, 120; antiscurvy measures 45-46, 144, foods **47,** 144
Second voyage (Antarctic, 1772-1775) 102-139; attacks by natives 130-131; crew 105, 107; homecoming 142; purpose 104; separation of ships 112, 118
Sedanka Island, Alaska 172, **176**
Seven Years War 8, 20, 22
Ship Cove, New Zealand 79, 113, 118, 138, 153-154
Ships: first voyage *see Endeavour;* second voyage *see Adventure; Resolution;* third voyage *see Discovery; Resolution*
Ship's discipline 6, 45, 46, 72, 98, 116, 161
Simcoe, John 20, 23
Sinoto, Yosihiko H. 161-162
Skottowe, Thomas 12, 20, 36
Society Islands, French Polynesia 154, 161; naming 68; *see also* Bora Bora; Huahine; Moorea; Tahiti
Solander, Daniel 36, 56, 64, 69, 82, 83, 104, 210
South America 44-45, 138
Southern continent *see Terra Australis Incognita*

Sparrman, Anders 112, 116, 118, 129
Spöring, H. D. 36, 57, 99
Square-rigger, modern copy of 18th-century ship **65**
Staithes, England **10-11,** 12-13, **18-19**
Statues: Easter Island **102-103, 112, 113-115,** 119, 120; Hawaii **202, 203;** Hiva Oa, Marquesas 129
Stearn, William T. 83, **91**
Stephens, Philip 35, 99, 142, 210
Stewart, Mike 74, **75**
Storms 118, 162, 168, *168,* 169, 178, 185, 193
Stranding: Great Barrier Reef 87, *94*
Stubenrauch, Fred 82
Surveying 23; *see also* Charting

Tahiti, French Polynesia 9, 34, *34,* **37,** 41, **42-43,** 46, 61, 64, 116, 160-161; map **45;** observatory *46,* 53
Tahitians 46, 56, 116, 178; chiefs 56, 64, 116, 160-161; fishermen **42;** human sacrifice 160; infanticide *49,* 57; polygyny **49;** traditional dishes 56; women **38-39, 48-49**
Tahuatu (island), Marquesas *122,* 129
Takutea (island), Cook Islands 154
Tanna (island), New Hebrides **124-128,** 131-132; circumcision rites **133-137**
Tapa cloth *see* Bark cloth
Tapuaetai (islet), Aitutaki (island), Cook Islands **145**
Taro **52, 53,** 68-69, 73, 132, **159,** 162
Tasman, Abel Janszoon 69, 72, 74, 79, *81,* 130
Tasmania, Australia 9, 113, 153
Tattoos 61, **72, 73,** 118, 129, **161**
Taunton, England 9, 26
Tautira (district), Tahiti 61, 116
Te Horeta (Maori) 74
Tepa, Leon **52, 53,** 68-69
Terra Australis Incognita (supposed southern continent): search for 34-35, 45, *45,* 69, 72, 104, 113, 119, 138
Thievery 46, 56, 116, 120, 130, 154, 155, 161, 202, 207
Third voyage (North Pacific, 1776-1780) 140-211; Cook's change in attitude 155, 192-193, *193;* crew 143, discontent 192-193, *193;* northernmost landfall 177, *187;* purpose 143, 168, 169; Tongan plot 155, *160*
Tierra del Fuego, South America **35,** 44, 45, 138
Tiwaka River, New Caledonia **139**
Tolaga Bay, New Zealand 73
Tonga, South Pacific 69, **108,** 117, 130, 154, 155, **156-159,** 160, 161; map **155**
Tongans 117-118, 130, **156-158;** chiefs 117; paramount chief 160; dance **110, 111,** 160, **160;** feast **158-159,** 160; nobles' plot 155, *160*
Tongatapu (island), Tonga **108,** 160
Trading 53, 57, 73, 117, 120, 129, 130, 131, 154, 155, 162, 168-169,

192; fur trade *173, 178, 178,* 184
Trevenen, James 169, 177, 178, 185, 195, 210
Tribulation, Cape, Australia 87, **92-93,** *94*
Tuamotu Archipelago, French Polynesia **32-33,** 46, 116, 129
Tubuai (island), Austral Islands **62-63**
Tupaia (Tahitian priest) 56, 68, 69, 73, 79, 83, 99, 116
Tupou, Fifita 155, 160
Turnagain, Cape, New Zealand *69,* 73
Turnagain Arm, Cook Inlet, Alaska **174-175**

Umiaks 173, **180-181, 183**
Unalaska Island, Alaska 172, 178, **178, 179,** 184, 185

Vancouver, George 143, 168, 202
Vancouver Island, Canada 168-169, **172, 173**
Vanuatu *see* former name, New Hebrides
Villiers, Alan 13, 45, 110

Waimea Bay, Kauai (island), Hawaii 162, **162,** 202
Wales, William 107, 129
Wales, Alaska 173
Walker, John 13, 20, 36, 104, 105, 142, 210
Wallis, Samuel 34, 41, 53, 56, 107
Walruses 173, 177, **182;** ivory 177, **184**
Water, fresh 79, 112, 132, 153, 155, 178; from icebergs *105,* 110
Waterspouts **108-109,** 113
Watts, John 192
Webb, Clement 64
Webber, John 143, **173;** paintings by **143, 172, 173, 182**
Whitby, England 13, 17, **21,** 104, 105
Williamson, John 155, 162
Wolfe, James 23

Yasur Volcano, Tanna (island), New Hebrides **124-125,** 131
Yorkshire, England 9, 12, **14-17;** *see also* Staithes *and* Whitby
Young, Nicholas 72, 73, 99

Zoology *56,* 72, 87, *87,* 90, *107,* 113, *148,* 154, *171;* paintings *57,* 107

Library of Congress CIP Data

Gray, William R. 1946-
  Voyages to paradise.
  Bibliography: p.
  Includes index.
  1. Cook, James, 1728-1799. 2. Explorers—England—Biography. I. Gahan, Gordon W. II. National Geographic Society, Washington, D. C. Special Publications Division. III. Title.
G246.C7G68   910'.92'4 [B]   78-21187
ISBN 0-87044-284-8 (regular binding)
ISBN 0-87044-289-9 (library binding)

Composition for VOYAGES TO PARADISE by Composition Systems Inc., Arlington, Va., and National Geographic Photographic Services (calligraphy and index pages). Printed and bound by Holladay-Tyler Printing Corp., Rockville, Md. Color separations by The Beck Engraving Co., Philadelphia, Pa.; The Lanman Companies, Washington, D. C.; National Bickford Graphics, Inc., Providence, R.I.; Progressive Color Corp., Rockville, Md.